Myth & Mirage

Myth & Mirage

Inland Southern California, Birthplace of the Spanish Colonial Revival

Introduction by
Catherine Gudis

Essays by
Aaron Betsky
H. Vincent Moses and Catherine Whitmore
Lindsey Rossi
Carolyn Schutten
Patricia A. Morton
Ronald Ellis (Interview)
Susan Straight

Photographs by
Douglas McCulloh

Edited by
Senior Coordinating Editors H. Vincent Moses and Catherine Whitmore
and Editor Ai Kelley

RIVERSIDE ART MUSEUM *Riverside, California*
PUBLISHED WITH THE ASSISTANCE OF THE GETTY FOUNDATION

CONTENTS

8   Acknowledgments

11 INTRODUCTION *Catherine Gudis*
    The Aesthetics of Amnesia
    Mapping Spanish Colonial Revival and Racial Geographies of Inland Southern California

1
    Aaron Betsky
    The Wall and Its Dissolution
27 The Spanish Colonial Revival from Style to Vernacular

2
    H. Vincent Moses and Catherine Whitmore
43 **Castillos, Iglesias y Casas**
    Constructing the Spanish Colonial Revival in the Inland Empire, 1895–1935

3
    Lindsey Rossi
    History by Design
77 Staging the Spanish Colonial Revival in the Inland Empire

101 SIDEBAR *H. Vincent Moses and Catherine Whitmore*
    Building the Mission Inn

4
    Carolyn Schutten
    Voids of the Aggregate
111 Materializing Ethnic Mexicans in Mission Revival and Spanish Colonial Revival in the Inland Empire

5
    Patricia A. Morton
    Postwar Spanish Colonial
133 **Revival Architecture in Inland Southern California**
    From Mission Inn to Taco Bell

6
    H. Vincent Moses and Catherine Whitmore
153 **The Spanish Colonial Revival at California Baptist University**
    Interview with Dr. Ron Ellis, President

7
    Susan Straight
    Bougainvillea, Lavandería,
163 **Cemetery, and Cross**

177 Selected Reading List
179 Biographies
181 Index

ACKNOWLEDGMENTS

The idea of presenting the extraordinary contributions to the development of the Spanish Colonial Revival in the Inland Empire was inspired by the suggestions of architects Michael Burch and Diane Wilk of Michael Burch Architects; thanks to the scholarly expertise of this project's contributors, the Riverside Art Museum (RAM) and Curator Lindsey Rossi are proud to have been able to expand the scope of the project to include reflections on the complex cultural implications of the style and its construction throughout the region. Many organizations and individuals were instrumental in bringing this project to fruition.

We extend profound gratitude to the Getty Foundation for believing in the importance of this book and exhibition, and for including *Myth & Mirage* among the other esteemed Pacific Standard Time: LA/LA exhibitions. Without their generous grant contributions, neither would have been possible. They provided guidance throughout the lead up to the exhibition opening in September of 2017 and the release of the catalog.

For her early guidance at the onset of this project, we wish to extend deepest gratitude to Mary MacNaughton at the Ruth Chandler Williamson Gallery at Scripps College.

Many thanks to Sheila Bergman, Tyler Stallings, and our colleagues at the University of California Riverside's ArtsBlock for sharing this opportunity with us and collaborating on programming and marketing. Downtown Riverside is rich in artistic programming and this joint collaboration is an incredible opportunity to share this with the larger Southern California region.

Many thanks are due to the authors who have contributed a high level of scholarship to this catalog and whose essays helped to establish the framework for the exhibition: Aaron Betsky, H. Vincent Moses and Catherine Whitmore, Lindsey Rossi, Carolyn Schutten, Patricia Morton, and Susan Straight. We thank Professor Catherine Gudis, Chairperson of the Graduate Program in Public History at UC Riverside, for providing her erudite introduction to this catalog.

This book and the exhibition are greatly indebted to Moses and Whitmore who assumed the coordination of this exhibition catalog and served as diplomatic and decisive liaisons throughout the editorial process and in accordance with the exhibition plan. Both also served as de facto consultants for virtually all manner of local history inquiries for nearly all of this book's authors. Their resource recommendations and expertise have been crucial to the success of both the catalog and exhibition. We thank Tish O'Connor for her keen insights and feedback on the first drafts of the essays in this volume. We are indebted to our expert Publication Review Panel readers who offered valuable recommendations to our authors. Thanks to Dr. Jonathan Tavares at the Art Institute of Chicago and Courtney Stewart at the Metropolitan Museum of Art for their specialized expertise and peer reviews of Rossi's design and decorative arts oriented essay.

The esteemed artist photographer Douglas McCulloh tirelessly provided publication photography upon request for several authors, while also developing an ingenious exhibition design in coordination with the curatorial requests of Rossi. We are extremely grateful to have had access to his inspired compositions.

The entire RAM *Myth & Mirage* team wishes to express deep gratitude to Ed Marquand, Adrian Lucia, Donna Wingate, Melissa Duffes, Tom Eykemans, and Ryan Polich at Lucia|Marquand for their unwavering professionalism in creating an arresting publication to seamlessly accompany the exhibition.

The authors commend the assistance of many local organizations and individuals, including: Dr. Ronald L. Ellis, the President of Cal Baptist University; Karen Raines, Curator of History, and Steve Spiller, Executive Director, of the Mission Inn Museum; Kevin Hallaran and the Riverside Metropolitan Museum Archive; Ruth McCormick and the Riverside Public Library Local History Resource Center; Nathan Gonzales and the A. K. Smiley Public Library History Room; the Palm Springs Historical Society; Corona Public Library History Room; the University of California Riverside Special Collections & Archives; Tina McIntyre, Vito Thornton, and Cal Portland Historical Archives Collection, as well as to author Chuck Wilson, and Gary Thornberry and Gene Juarez for sharing their personal stories of life in a cement factory; and Sandra Schmitz for her research assistance. Special thanks to Laura Sorvetti, Library Services Specialist, and Jessica Holada, Director of Special Collections and Archives at the Robert E. Kennedy Library at Cal Poly San Luis Obispo for repeated access to the Julia Morgan archives. Thanks also to Sunset Magazine and former Editor in Chief Peggy Northrop for the invitation to explore their historic editions and the magazine's former home in Menlo Park.

The generosity of lenders for the exhibition throughout Southern California has been extraordinary, together with the efforts and courtesies of their staff: The Mission Inn Museum (Steve Spiller, Karen Raines); Riverside Metropolitan Museum (Kevin Hallaran); Riverside Public Library Local History Archive (Ruth McCormick); California Baptist University Queenie Simmons Archive (Vi Estel, Tammy Pettit); Special Collections at Cal Poly San Luis Obispo's Robert E. Kennedy Library (Jessica Holada, Laura Sorvetti); Bowers Museum (Peter Keller, Victoria Gerard, Katie Hess); Art, Design and Architecture Museum at the University of California Santa Barbara (Jocelyn Gibbs, Julia Larson); Automobile Club of Southern California (Matthew Roth), Bauer Pottery (Janek Boniecki); Jean Aklufi; Kathie Allavie; Sue Johnson; H. Vincent Moses and Catherine Whitmore; Todd Wingate; Carolyn Schutten; and Valerie Found.

Many thanks to our Board of Trustees for prioritizing exhibitions that are relevant to our community. This exhibition represents a large step forward for the museum and four years of collaborative work.

Many at RAM were instrumental in bringing this project to fruition. We are thankful to Ai Kelley for her unflagging efforts marketing the exhibition and catalog to the diverse population of Southern California and beyond; to Todd Wingate and Allen Morton for their ingenuity and industry as the exhibition design took shape and for handling the logistics of this complex undertaking; to Katie Hernandez and Brightie Dunn for organizing docent-led tours of the exhibition; to Caryn Marsella for her efforts implementing educational programs; to Eric Romero and artist Cynthia Huerta who engaged the community at Our Lady of Guadalupe and gave insight into their contemporary lives and their use of this building featured in this project; and to Valerie Found for her tireless fundraising efforts on behalf of the museum.

Additional acknowledgement must also be extended to RAM's Art Alliance, Tim Burgess of Burgess Moving and Storage, the Old Riverside Foundation, and the City of Riverside Historic Preservation Division.

Drew Oberjuerge, Executive Director
of the Riverside Art Museum, and
Lindsey Rossi, guest curator

INTRODUCTION

Catherine Gudis

The Aesthetics of Amnesia

Mapping Spanish Colonial Revival and Racial Geographies of Inland Southern California

In a historic moment in which social rights movements have found new blood and mobilized masses of people of color and those of varying backgrounds and political persuasions, commonly called to action in the digital and physical arenas, we are overdue in staking claims to the ways in which race is spatially inscribed in the built environment, and embedded in the architectural fabric and regional identity of Southern California.[1] All too often, architectural style seems to get a "free pass"— an exemption from critique in analyses of racialized landscapes— even while we've interrogated the urban policies and planning that guide the use and construction of buildings, the banks that mortgage them, and the neighbors who bought homes with covenants restricting ownership by African Americans, Jews, and other ethnic groups. We know that style matters, and that it is an expression of the sociocultural moment and political economy of its creation and recreation, as well as the individual will or agency of a range of actors, from owners and architects to inhabitants and tastemakers (critics, real estate developers, boosters, preservationists). Yet the results of these interrogations and the larger context of how style matters often do not become part of the public articulations about a place, and architecture becomes a set of obdurate material facts—stylistic elements analyzed for their formal qualities and stylistic genealogies—denuded of politics, agency, and implication. How do we put architectural style into a racial geography that exceeds this

FIGURE 1 *Campanario* built during New Deal-era reconstruction (1926–37) of the San Bernardino Asistencia (outpost of Mission San Gabriel), with shadow of El Camino Real commemorative bell. Redlands, 2017

denigration of sociohistorical significance, and acknowledges our own roles in producing meaning, in the current era and in deciphering what style might have meant to different people in different eras and localities?

If ever there was a style begging for such a treatment it is Spanish Colonial Revival (SCR),[2] whose racial and political genesis and implications span the globe, including migratory routes between California and Mexico (where it came in the 1920s to be called *Colonial Californiano*), and whose ubiquity in Southern California has all but eviscerated our ability to see it much less to see it as mattering.[3] We have for the most part forgotten how it came to define our region, and why (figs. 1–3). Surely we have not yet sufficiently queried why SCR has had such staying power, still part of the design guidelines for countless communities and the dominant expression of place identity from Santa Barbara to Beverly Hills and from Redlands to Riverside.[4] Why has it been so productive, especially when the biggest population growth and architectural innovations in California history are of a different era and style, of the modern, born out of mid-twentieth-century military technologies, industries, and automobility?[5] How might the fantasies of racial conquest embodied by SCR explain its persistence and its participation in power dynamics that affect ideas of exclusion and belonging? What might be the ongoing implications and narrative possibilities of SCR in broadcasting these dynamics of race, space, and place?

The current volume puts SCR back into view, to remind us where it came from, how it is part of the fabric of everyday life in Southern California, and that it matters. It does so in part by training our eyes on the aesthetic form of SCR as crystallized particularly in the Inland Southern California region. In what

FIGURE 2 Walgreens, Riverside, 2017

FIGURE 3 Tower and main building, El Mirador Hotel (now Desert Regional Medical Center), reconstructed in 1991 based on 1928 original, Palm Springs, 2017

follows, SCR is analytically and photographically dissected, taken apart and then put back together again so that we see the revival style anew and for ourselves. In doing so, it reveals SCR as nearly everywhere around us in different classes and scales of residential communities, civic centers, and commercial strips, mass produced in built form and as imagery layered through and constitutive of our mythologies of the region (figs. 4–7). How the style connects past and present, over and again, across different temporalities, and how the persistently reborn mythos of SCR has functioned in the region are among the questions addressed in what follows. The essays, contemporary photographs, and historical source materials serve as an opening gambit—and provocation—to probe further and more deeply, and they provide us with tools to do so.

Though the architectural story of SCR expressed in these pages is one of presence (what we can see on the landscape today as documents of the past), it is also about absence and erasure. Indeed, an aesthetic of amnesia is endemic to SCR, and scholars of California history have long addressed its fabrication of a "fantasy" Spanish past that masks a history of conflict and conquest, and its dominant feature of whitewashed adobe as embodying the larger deracination or whitening of historical narratives about California that seek to dispossess Native and Mexican people of their history and stake to the land.[6] Developed as an architectural idiom in the same decades as the first exclusionary acts of immigration policy were articulated (from Chinese exclusion in 1882 to the establishment of the border patrol and immigration quotas in 1924 to deportation/

FIGURE 4 Model of Santa Fe Depot (in Santa Fe Depot), San Bernardino, 2017

13

repatriation of Mexican American citizens in the 1930s) and as concerns about racial purity were exercised by California's growing eugenics movement, SCR's stucco expanses and lime-coated walls offered a visual antiseptic to cultural anxieties around immigration, heterogeneity, and miscegenation. It asserted ideas about belonging, by virtue of who was cast as foreign or "other" to the narrative, perhaps as a means to foreclose the possibility that anyone besides those Anglos of the dominant classes who did the commissioning and promoting of SCR might lay claim to their rights to the land and landscape.

Progressive journalist, activist lawyer, and historian Carey McWilliams famously articulated the racial schema, suppression of historic violence, and erasure of the Indian and Mexican past that was part of the region's "Mission Myth" and "Fantasy Heritage" in his still-influential *Southern California: An Island on the Land* (1946) and what is considered the first book-length history of the Chicano experience, *Notes from Mexico* (1948). In these volumes he trenchantly critiques the American manufacture of a "curious legend" regarding the mission era, as a time of benevolent paternalistic friars and their grateful Indian friends. The narrative for the period after secularization occluded the Mexican presence, and instead depicted the region as romantically populated by Spanish residents, "all members of one big happy guitar-twanging family, [who] danced the fandango and lived out days of beautiful indolence in lands of the sun that expand the soul" (figs. 8–10).[7] McWilliams dismantles and

FIGURE 5 La Plaza, Palm Springs, 2017

FIGURE 6 Donut Avenue, Chino, 2017

FIGURE 7 Remington Estates, Chino, 2017

FIGURE 8 Promotional brochure, California Parlor Car Tours Company, ca. 1930

FIGURE 9 "Spanish" entertainment, Old Adobe of Mission Inn, Riverside, ca. 1925

FIGURE 10 Fandango scene in courtyard, constructed of oranges, Riverside exhibit, National Orange Show, San Bernardino, ca. 1927

exposes the historical fabrications at play, coolly offering "the facts," that "Franciscan padres eliminated Indians with the effectiveness of Nazis operating concentration camps" and the function of SCR's stylistic inventions: "to deprive the Mexicans of their heritage and to keep them in their place."[8] Within the racial order he delineates (and that scholar Roberto Ramón Lint Sagarena elaborates on), Mexicans were cast as "perpetually foreign" in the United States, "happily forgotten" by Anglo promoters, whom he castigates as instead heralding "the restored Mission [as] a much better, a less embarrassing, symbol of the past than the Mexican field worker or the ragamuffin *pachucos* of Los Angeles."[9]

Scholars since McWilliams have built upon his critique in nuanced ways, to consider how and why SCR privileged an Iberian past and fabricated a European lineage for California. They suggest this served several functions. It offered Southern California a "usable past" that was European and "civilized," matching its Mediterranean climate and contradicting naysayers who saw the region as still a western cow town and rough frontier.[10] American hispanophilia also created continuities between the Spanish colonial enterprise and American expansion, naturalizing and justifying Anglos as the heirs to that Spanish enterprise, rather than the Native Americans, Californios, and ethnic Mexicans who lost out in the aftermath of the Mexican American War, no matter what side they had chosen. The version of the Spanish past fabricated in the last decades of the nineteenth century, in other words, helped create California as Anglo, distancing it from the Mexican and Indian past and present; as hundreds of thousands of Mexicans entered California in the aftermath of the Mexican Revolution in the first decades of the twentieth century, a reinvigorated Spanish Fantasy Past again offered a comforting myth and means of reasserting racial distinctions and superiority (fig. 11).[11]

The Mission Myth and Fantasy Past both invented traditions and enabled collective amnesia, surely in part derived from willful refusal, then and now, by promoters and preservationists as well as art historians and architects, to acknowledge historical complexity—especially as pertains to race. Lost with this amnesia is the social cost of SCR. Buildings and

FIGURE 11 Promotional pamphlet describing various towns in Los Angeles County, 1925

urban spaces often seem to be race-neutral, as George Lipsitz explains in *How Racism Takes Place*, belying the historical processes of their construction.[12] This is particularly true when one considers even the nomenclature "Spanish Colonial Revival," a term that now rolls off the tongue like any stylistic descriptor, as if this too is race neutral. Yet the very phrase valorizes colonial practices, which we know are far from race neutral and, moreover, are freighted with ideological weight and the legacies of empire. Though deployed in the aesthetic realm and perhaps in a postcolonial era, even "revived" colonial practices still carry the initial colonial weight. They continue processes of settler colonialism, in which native elimination and immigrant exclusion are structured by racial fantasies of conquest and white supremacy. These were as much a part of Spanish colonial practices and, after 1848, American nationalist practices as they are about walls, borders, and deportations demanded by xenophobes today. Which is to say, *style* still serves symbolically in structuring space according to hierarchies of race.

Some might argue that an architectural revival can't possibly continue a process of domination and dispossession already seen as completed in earlier eras, and that it is overdrawn or just too strident to connect the violence of the past with the violence of the present. Besides, what would be the point of returning to memory the less romantic roots of SCR in the missions that were established in the Americas and in California from 1769 to 1823? It can't be good for business to remember the violence that colonization entailed. Who wants to buy a building of adobe walls and courtyard organization referring to missions stained with the blood and sweat of the native people who labored and died in huge numbers in such spaces? Or who rose up against colonial exertions of power, or quietly resisted—continuing cultural practices in those same courtyards where punishment would be meted for such acts? California Indian mortality was expedited by the mission architecture itself, which concentrated large numbers of people in enclosed spaces as petri dishes for disease. Many indigenous inhabitants had been shackled there and others—particularly women and children—locked in, ostensibly for protection from illicit sexual encounters (contradicted by larger incidence of rape and violence). The walls of SCR dividing interior and exterior space, even if only formal gestures to this past, remain laden with this history of social control, carcerality, and outright violence.[13]

When formal stylistic devices from the missions were appropriated to create spaces of privilege and privacy for the middle and upper classes in the twentieth century, it is no surprise that such remembrances weren't passed along with the rest of the marketing materials. Figuring that not everyone would savor sipping sangria in such a courtyard, clever promoters instead advertised "hacienda styled" living, with noble dons and fandango to round out the picture. Lost in translation is that "hacienda" means *plantation*, indicating a formal kit of architectural parts (exterior arcades, connected blocks of interior space, whitewashed walls, courtyards) for compounds that combined communal living and forced labor. These existed throughout the Americas and beyond; in California the labor of Native Americans was forced through physical violence and modes of control in the mission era, and enslaved or indentured in the American era.[14] The nineteenth and early twentieth-century SCR fantasy creation of ornamental gardens in what were originally utilitarian courtyards masks their historical function and violent past, transforming them into sites of popular leisure and tourism (fig. 12).[15] What drew Native people to the courtyards of these missions and hacienda plantations (some of which are said to still have the posts that were used for whipping recalcitrant or unbending neophytes[16]) of course remains unsaid in the marketing of both the missions and the wider SCR style today. After all, those reasons for Indians being drawn to the missions and to secular plantations—loss of one's land, wholesale ecological destruction due to overexploitation of resources, decimation of subsistence economies, natural disaster (also an effect of overuse), and coercion by soldiers and Franciscans—are hardly topics to chat about over tapas in one's courtyard garden.[17]

In its historical erasure and the supplanting of courtyard whipping posts for gardens, SCR creates a tabula rasa. It wipes clean culpability and human cost. In so forgetting, SCR reproduces the systemic dispossession of indigenous people in the colonial era once again in the present day by shearing them from the record. Also annihilated is the record of Indian resistance and insurgency in the face of domination. Historically, "no mission escaped uprising from within or attacks from outside by communities of the imprisoned along with escapees."[18] Notably, Kumeyaay destroyed Mission San Diego in 1775; Gabrielino attacked Mission San Gabriel in 1785; and Chumash rebelled at Missions Santa Barbara, La Purísima Concepcíon, and Santa Inés in 1824.[19] It should surprise no one that SCR might not inspire comfort or nostalgia for Native Americans. Celebrations of SCR that fail to acknowledge the Indians who are still here, as living proof of these histories of violence, or that fail to give voice to the absences in the invented traditions at the root of SCR and acknowledge the power relations of spatial and architectural production, reproduce offenses from the past as they continue histories of forgetting.

FIGURE 12 Garden courtyard, Mission San Juan Capistrano, 1960s (from Kodachrome slide)

Preservationists in recent years (and I count myself among them) talk about buildings as containers for both historical memory and for contemporary meaning. Structures act as witnesses and material cues to the kinds of stories that ultimately make places matter, offering infill to material sites through associations, symbolism, and all the intangibles of social value that are individually and collectively construed.[20] What is erased, written over, or ignored, including the histories of violence, coercion, resistance, and most of all, mass mortality, are among the associations and symbolisms of SCR that remain for scholars, storytellers, and culture bearers to address, in order to counter the continuing historical annihilation of Mexican and Native American people, and to properly reflect their rights to space and place in California. To the informed reader of both this volume and of our California landscape, SCR necessarily marks these rights as well. The SCR "containers" hold these meanings and memories, too, especially when we bring to light what has fallen away from public understanding. Seen this way, the present volume intends itself not so much as a final word about the SCR style of California, but as a call for more—more narrative voices that exhume the past and add to its present, and that remap the racial geographies of SCR.

The first half of the book tours us through the SCR landscape, with particular focus on Inland Southern California. In the chapter that follows, Aaron Betsky's "The Wall and its Dissolution" synthesizes the style and its genealogy, offering a phenomenological accounting or feel for the style, bringing us into the experience of the buildings through their elements. At the start of his essay, Betsky makes the key claim about the white patrimony of SCR: "The SCR, above all else, marks the place of the wealthy and powerful, or those who would aspire to be so. It defends and represents what they have made for and of themselves." Furthermore, as he describes, the middle class "newly arrived in both place and in an economic sense" could make, through SCR, "a space for itself that was comfortable and comforting." As Betsky shows, the establishment of the architectural style of SCR offered Anglo transplants to the region social cohesion by connecting them through a shared romance with the past. He concludes most evocatively that SCR enabled a "middle-class paradise for Anglo settlers."

The "comfort" derived from references to a Spanish Catholic past remain curious for an era in which there was deep Papal distrust and Anglo Protestant dominance, as H. Vincent Moses and Catherine Whitmore indicate in their essay, "Castillos, Iglesias y Casas." This central contradiction of SCR was assuaged by the cast of characters that McWilliams calls the "troubadors" of "magical improvisation," and which Moses/Whitmore as well as Lindsey Rossi ("History by Design") address: mythmakers from the 1880s through the 1930s whose focus was decidedly secular, and aimed to distance "building forms from their church affiliation" and "to create a bourgeois Spanish Arcadia." Such figures included Charles Fletcher Lummis, who worked passionately to rescue Junípero Serra and Spanish Colonialism from what he saw as America's blinding Anglo Saxonism, and who founded the Landmarks Club and edited *Land of Sunshine* to save both the missions and other "relics of the heroic past"; Helen Hunt Jackson, who penned tracts to popularize the Franciscans as self-sacrificing civilizers of the land and its peoples, even as she sought to draw attention to the plight of Native Americans displaced from their land, stripped of their rights, and relegated to poverty under American rule; and John McGroarty, lawyer and state poet laureate whose *Mission Play* was seen by millions and who was a strong advocate of the California Studies curriculum that today still brings fourth graders to missions or to make models of them.[21]

Moses/Whitmore explain that "the eclectic fantasy" of SCR and its mission precedents "was nearly all pure myth and symbol, mixed liberally in a cultural cauldron to produce architecture embodying the idea of a new and perfected Mediterranean." Rich stylistic details run through this essay, touring us virtually through SCR's permutations, lavishing attention on its various facets to bring us into its encompassing architectural orbit. The focus on the Mediterranean flavor that served as a promotional tool to sell Southern California to those from the East and Midwest who thought the place was uncivilized and without history comes to bear here. Moreover, the authors show how leading citizens and their architects chose SCR to denote both wealth and empire, when citrus was "king" (from which the name Inland Empire also derives) and not long after the city of Riverside could boast in 1895 of the highest per capita income in the nation.[22]

Moses/Whitmore illuminate this further in their illustrated sidebar delineation of "Building the Mission Inn," and join Lindsey Rossi to highlight the Inland Empire's pre-eminence in staging fantasies of the past in grand and lavish terms, as a non-stop Spanish Revival Oz, to quote state historian Kevin Starr, constructed in the center of downtown Riverside over the first three decades of the 1900s. They rightfully emphasize the Mission Inn as the *ur*-architecture of SCR, though, as Moses/Whitmore explain, the set of structures started life as the decidedly Anglo (and remarkably humble) Glenwood Cottages, until proprietor Frank Miller gained support and funding from railroad magnate Henry E. Huntington and hired Los Angeles architect Arthur Benton, who "rendered the new hotel in an audacious Mission Revival." It kept growing in audacity with the help of others in the decades that followed. Today it still serves as a phantasmagoria of the colonial, its convergence of historical reality and illusion enacted through a spectacular assemblage of Mission and Spanish Colonial-era objects and exoticized architectural forms.[23]

Rossi shows in descriptive terms the Mission Inn as centerpiece for touristic itineraries and mythmaking function, which re-directed attention to the inland region and cast its lineage as related to Franciscans and conquistadors though nary a mission could be found within some fifty miles (where San Gabriel Mission was located to the east, and San Juan Capistrano Mission to the south). Geographical reality didn't stop Mission Inn proprietor Frank Miller from donning Franciscan robes and strolling the grounds as Father Serra, or from incessantly adding arcades, bell towers, patios, courtyards, a cathedral, and, in the Cloister Wing, a warren of underground galleries called the Catacombs that housed a small portion of his vast Spanish art collection. When visitors tired of the Mission Inn, they could hop on Miller's trolley and visit the constellation of SCR buildings comprising Sherman Institute. One of the last Indian boarding schools to be built by the federal government, as Rossi describes, native youth in the first half of the twentieth century were forcibly assimilated (stripped of their language and culture, forced to stay, sent out to perform servile labor, and trained to join the ranks of the underclass) under the watchful eye of administrators and the touristic gaze of well-coiffed and likely well-meaning Anglo visitors.[24] Today, only one of the original SCR Sherman buildings stand and the school operates under Native American supervision. The shadow of the missions is cast there and across the region still, as the dominant logo for the City of Riverside to this day (as Rossi describes) is the Raincross, derived from a bell (purportedly used by Serra) and a cross from Miller's immense collection.

A nearby campus of a different sort and era is California Baptist University, whose original buildings were designed by noted local SCR architect Henry Jekel (ca. 1925–38) as a home for aging members of the Neighbors of Woodcraft during the heyday in America of fraternal societies. As Ron Ellis, President of CBU, explains in the interview in this volume, SCR has become a fundamental part of the sense of place and gracious sensibility that the university conveys. In the 1990s, Ellis spearheaded a campus-building program that tactically applied SCR to new construction and thoughtfully preserved the original buildings. The interview aptly connects past and present uses of SCR, and the possibilities for transitions between them. The buildings at CBU make apparent the ongoing malleability of SCR to transcend time, original function, and religious reference.

The second half of the book more purposely and evocatively considers the implications of SCR, particularly in terms of labor, the postmodern era, and the social landscape, and is a crucial intervention into the existing literature of SCR. Carolyn Schutten's poetic title "Voids of the Aggregate" references the spaces between the aggregates (sand, gravel, etc.) used in the production of concrete, which, arguably, play a role in the strengthening and durability of this dominant building material of SCR. She uses this as a metaphor for gaps and erasures in the historical record, explaining, "Like the voids in concrete, the role of ethnic Mexicans in Mission Revival and SCR architecture has been largely omitted from the historical record, yet their labor and the production of building materials have played an integral part in the architectural history of the Inland Empire and Southern California." Importantly, this essay does just what is needed for so many architectural histories: to dissect who did the work in constructing the buildings. In this case, the obligation is doubly important, given the ways in which the very population groups who are erased in SCR's mythologizing are the Mexican and Native American people who labored to construct its inventions. She identifies both the industrial aspects of the style (cement and brick industries were key economic tools contributing to the region's development and growth) and the people behind it, with incredible historic photos of a largely multiracial labor force. The essay addresses ways to repopulate and shift the narrative of the otherwise deracinated landscape of SCR. Concluding with homes built in the recent past in the SCR style for the barrio of Casa Blanca, Schutten demonstrates how the style circulates across class and race in later eras.

The final two essays in the book bring to a poignant and telling conclusion the long history of SCR in all its manifestations. In "Postwar Spanish Colonial Revival Architecture in Inland Southern California: From Mission Inn to Taco Bell," Patricia Morton offers a lineage for SCR after its apotheosis in the 1930s, tracing its hiatus in the immediate postwar decades and its rise again by the 1970s, when, on a sliding scale of taste it also fell from high to low. Its resurgence or re-revival, as she puts it, was congruent with the rise in Postmodernism, and a renewed interest in critical regionalism, the symbolic function of design, and uses of historical references. Marked by its application of transferable features (tile roofs, stucco, dome, arch, etc.) to just about any structure, neo-SCR—or "refried" SCR—serves as a short hand for the region's (mythical) Spanish character. Its rise, she suggests, was both to serve as an antidote to modernism's "universal ideal" and as part of a larger desire "for a more rooted existence and architectural style to match it." Yet this too is a veneer of connectedness or rootedness in the face of an endless expanse of otherwise bland lower- and middle-class suburban developments to spring up in the Inland Empire.

Susan Straight concludes the volume with a moving portrait of the social landscape of SCR from mission to Mission Inn, putting us in the place of those who have labored and lived in the shadows of both. We shiver with her and her daughters as they hear about the Native women who lived and died at Mission La Purísima Concepción and imagine ourselves at the troughs where countless women at Mission San Luis Rey labored at the outdoor lavandería, "marvelous in design, timeless and haunted as an ancient Roman ruin."

Straight breathes life into the mission past and its Spanish colonial revival, helping us to rethink and recalibrate "the history we're given, about our romantic obsessions with Spanish Colonial Revival architecture and design, and the icons which are meant to represent an Edenic past." Real people from the past and present—Maria Juana de Los Angeles, Gordon Johnson, Yuliana Buenrostros, Roberto Loya—animate the Inland Empire she portrays, through multiple generations whose migrations and immigrations map the racial geographies of SCR and the region. Yet what Straight offers us is more than that. It is a chance for a reckoning and reconciliation with the past, an architectural revival of a different sort that brings back into view historical complexity and fixes our sights on the lives and legacies of those feared forgotten.

Notes

1   Such mobilization ranges from Black Lives Matter to Standing Rock to recent demonstrations in response to the contentious presidential election of 2016 and its aftermath, all of which have called for historical reckoning of the racial hierarchies of white supremacy structuring state violence—whether in regards to police violence, state-sanctioned support for corporate interests over indigenous rights, or "America First" nationalist rhetoric.

2   In his essay in this volume, Aaron Betsky notes that style is a challenging concept, since it suggests for some architects and critics that "the appearance of a building is both independent from its formal and structural properties, and is a question of choice." He defines the word, for his essay, in terms of "both an economic and social justification and a developed canon or coherent accumulation of elements." In the context of the present discussion, I would add that the visual coherence of SCR also serves as a specifically aesthetic form of cultural colonization, a projecting gaze, if you will, that subsumes and orders the landscape (an analogue to Mary Louise Pratt's *Imperial Eyes* [New York and London: Routledge, 1992]). Its commodification in turn can serve as a form of consumptive conquest. Or, as Phoebe Kropp writes, the Spanish "style appeared to grant an 'imagined proprietorship' over the region as a whole, its history and essence." Phoebe Kropp, *California Vieja: Culture and Memory in a Modern American Place* (Berkeley: University of California, 2006), 160.

3   The back and forth migrations and architectural transformations are narrated in the video by Jesse Lerner and Rubén Ortiz-Torres, *Frontierland/Fronterilandia* (1995), and synopsized in their related article, "Spanish Caprice," *Art Issues* (January/February 1996): 23–25. The point is further elaborated by Ana Elena Mallet, in an essay in the book accompanying Lerner and Ortiz-Torres' co-curated exhibition "Mex/L.A." She writes: California elite "'rediscovered' neo-Spanish architecture through studies on Mexican colonial art, and the style quickly returned (1921) to Mexico City, translated as *Colonial Californiano*" and was adopted as essentially "nationalist." Ana Elena Mallet, "Mexico City—L.A.: Design Dialogues," *Mex/L.A.: "Mexican" Modernism(s) in Los Angeles, 1930–1985* (Ostfildern, Germany: Hatje Cantz, and Long Beach: Museum of Latin American Art, 2012), 107.

In considering the mutually constituted racial geographies of Mexico and the U.S., Maria Saldaña-Portillo describes that "the ways in which national geographies are perceived, imagined, lived, and mapped are supremely racial," and that our contemporary geographies cannot be understood without investigation of the colonial and postcolonial placement and displacement of indigenous subjects in the landscape. Maria Saldaña-Portillo, *Indian Given: Racial Geographies across Mexico and the United States* (Durham, NC: Duke University Press, 2016), 6–7.

4   The Mission Revival style reached its apotheosis by the time of the 1915 Panama-California Exposition held in San Diego, when it became subsumed by the more lavish architectural umbrella of Spanish Colonial. See Matthew Bokovny, *The San Diego World's Fairs and Southwestern Memory, 1880–1940* (Albuquerque, NM: University of New Mexico Press, 2005). Within ten years of the San Diego fair, over fifteen cities would appropriate the Spanish style, including Ojai, Palos Verdes, Rancho Santa Fe (north of San Diego), Montecito, and San Clemente. Scores of individual contractors and tract developers built Spanish-styled houses, while homeowners were sold pattern books and kits to build their own. Elizabeth McMillian, *California Colonial: The Spanish and Rancho Revival Styles* (Atglen, PA: Schiffer Publishing, 2002), 32–35; Kropp, *California Vieja*, 159–206). When Santa Barbara was devastated by an earthquake in 1925, it was rebuilt in the Spanish theme, with the courthouse and mission centerpieces to the city; comprehensive zoning laws and, later, architectural review boards safeguarded elements of the comprehensive design to be followed on every level, down to mailboxes and trash cans. Roberto Ramón Lint Sagarena, "Building California's Past: Mission Revival Architecture and Regional Identity," *Journal of Urban History* 28 (May 2002).

5   See, in this volume, Aaron Betsky's essay for the ways in which primary elements of the wall and the box became important sources for modernism, and Patricia Morton's "Postwar Spanish Colonial Revival Architecture in Inland Southern California: From Mission Inn to Taco Bell" for postmodern quotations of the style. Both show the ways in which SCR might have had its roots in the first part of the twentieth century but infiltrates the twenty-first century, too.

6   As Deborah Miranda puts it: "Along with this visual mythology of adobe and red clay roof tiles comes the cultural storytelling that drains the missions of their brutal and bloody pasts for popular consumption." She amply illustrates this by darkly ironizing popular forms of mythologizing, including California's fourth-grade curriculum for mission-model building, which Susan Straight also discusses in personal terms in her essay in this volume. See Miranda, *Bad Indians: A Tribal Memoir* (Berkeley: Heyday, 2013), xvii. Also see William Deverell, *Whitewashed Adobe: The Rise of Los Angeles and the Remaking of its Mexican Past* (Berkeley: University of California Press, 2004); Kropp, *California Vieja*; Elizabeth Kryder-Reid, *California Mission Landscapes: Race, Memory, and the Politics of Heritage* (Minneapolis and London: University of Minnesota Press, 2016); William Alexander McClung, *Landscapes of Desire: Anglo Mythologies of Los Angeles* (Berkeley: University of California Press, 2000); and Roberto Ramón Sagarena, *Aztlán and Arcadia: Religion, Ethnicity, and the Creation of Place* (New York and London: New York University Press, 2014).

7   Carey McWilliams, *Southern California: An Island on the Land* (Santa Barbara and Salt Lake City: Peregrine Smith, Inc., 1946), 22, 70.

8   Ibid., 29; Carey McWilliams, *North from Mexico: The Spanish-Speaking People of the United States* (1948; 3rd ed. updated by Matt S. Meier and Alma M. García, 2016), 18.

9   McWilliams, *Southern California*, 83. In *Aztlán and Arcadia*, Sagarena advances McWilliams point by calling the fabrication a way of portraying ethnic Mexicans as familiar but nonetheless perpetually foreign in the United States (2, 6, 133–34). Also see Peter Richardson, *American Prophet: The Life & Work of Carey McWilliams* (Ann Arbor: University of Michigan Press, 2005), 142–43.

10   "Americans longed for a weighty history of their own," Sagarena explains, "that would pull their center of cultural gravity from the Old World to the New," and the crumbling coastal mission churches offered a counterpart to what one might visit on their Grand Tours of Europe (*Aztlán and Arcadia*, 19, 24).

11   Ibid., 126; William Deverell, "Privileging the Mission Over the Mexican: The Rise of Regional Identity in Southern California," in *Many Wests: Place, Culture and Regional Identity*, eds. David Wrobel and Michael Steiner (Kansas City: University of Kansas Press, 1997), 253; David G. Gutiérrez, *Walls and Mirrors: Mexican Americans, Mexican Immigrants, and the Politics of Ethnicity* (Berkeley and London: University of California Press, 1995), 57. As Gutiérrez details, "Historical migration statistics for this period are notoriously inaccurate" but he extrapolates that by 1920 Mexican nationals in the U.S. numbered 478,000 and in 1930, that number had "increased to at least 639,000." Gutiérrez, "An Historic Overview of Latino Immigration," *American Latinos and the Making of the United States: A Theme Study* (Washington, D.C: National Park Service, 2013), https://www.nps.gov/heritageinitiatives/latino/latinothemestudy/immigration.htm (accessed January 5, 2017).

12   George Lipsitz, *How Racism Takes Place* (Philadelphia: Temple University Press, 2011).

13   Scholars differ in their explanations for mission conditions and reasoning around violence, though commonly acknowledge high mortality, sexual violence, and punishment by flogging and shackling. Recent studies also highlight native resistance and agency in the face of catastrophic change. See, for instance, Virginia Bouvier, *Women and the Conquest of California, 1542–1840* (Tucson: University of Arizona Press, 2001); Miroslava Chávez-Garcia, *Negotiating Conquest: Gender and Power in California, 1770s to 1880s* (Tucson: University of Arizona Press, 2004); Steven Hackel, *Children of Coyote, Missionaries of Saint Francis: Indian-Spanish Relations in Colonial California, 1769–1850* (Durham: University of North Carolina Press, 2005) and *Junípero Serra* (New York: Hill and Wang, 2013); Robert Jackson and Edward Castillo, *Indians, Franciscans, and Spanish Colonization: The Impact of the Mission System on California Indians* (Albuquerque: University of New Mexico Press, 1996); and James A. Sandos, *Converting California: Indians and Franciscans in the Missions* (New Haven and London: Yale University Press, 2004). In her chapter, "Colonial Mission Landscapes," Kryder-Reid summarizes the uses of mission courtyards for purposes of social control by missionaries and as spaces of cultural resistance by Native Americans (*California Mission Landscapes*).

14   Benjamin Madley, *An American Genocide: The United States and the California Indian Catastrophe* (New Haven and London: Yale University Press, 2016), 36–39, 158–61, 286–87; Albert Hurtado, *Indian Survival on the California Frontier* (New Have: Yale University Press, 1988), 211.

15   Elizabeth Kryder-Reid enumerates this process in detail in *California Mission Landscapes*. She writes, "the gardens planted in the mission courtyards are not merely anachronistic but have become potent ideological spaces that naturalize California's valorized settler colonial narratives. The transformation of these sites of colonial conquest into physical and metaphoric gardens has perpetuated the marginalization of Indigenous agency and avoided confronting the contemporary consequences of colonialism" (4).

16   Roxanne Dunbar-Ortiz, *An Indigenous Peoples' History of the United States* (Boston: Beacon Press, 2014), 128.

17   Wholesale changes to the environment due to the importation of European agricultural models along with a prolonged drought drew many Native Americans to the missions, where the padres promised them sustenance and gave them indentured servitude instead. Also see Deana Dartt-Newton and Jon M. Erlandson, "Little Choice for the

Chumash: Colonialism, Cattle and Coercion in Mission Period California," *American Indian Quarterly* 30 (Summer/Autumn 2006): 416–30.

18   Dunbar-Ortiz, *Indigenous Peoples' History*, 129.

19   Lisabeth Haas, *Saints and Citizens: Indigenous Histories of Colonial Missions and Mexican California* (Berkeley: University of California Press, 2013), 54–55, 120–21, 116–39 (chapter 4, on the Chumash War), 155; Hackel, *Children of Coyote*, 259–61 and 263–66; Hackel, "Sources of Rebellion: Indian Testimony and the Mission San Gabriel Uprising of 1785," *Ethnohistory* 50, no. 4 (2003): 643–69; Richard Carrico, "Sociopolitical Aspects of the 1775 Revolt," *Journal of San Diego History* 43 (Summer 1997): 142–57; and, more generally, George Harwood Phillips, *Chiefs and Challengers: Indian Resistance and Cooperation in Southern California, 1769–1906*, 2nd ed. (Norman: University of Oklahoma Press, 2014).

20   Ned Kaufman, *Race, Place, and Story* (New York: Routledge, 2009), 3, 9, 70.

21   The "troubadors" of "magical improvisation" are described in Carey McWilliams, *Southern California*, 21. On California Studies curriculum and the fourth-grade mission project, see Zevi Gutfreund, "Standing Up to Sugar Cubes: The Contest Over Ethnic Identity in California's Fourth-Grade Mission Curriculum," *Southern California Quarterly* 92 (Summer 2010): 161–97; Phoebe Kropp, "Sugar Cube Missions: Bringing the Spanish Past into the California Classroom," paper presented at the 94th Organization of American Historians Annual Meeting, Los Angeles, April 2001.

22   H. Vincent Moses, "Machines in the Garden: A Citrus Monopoly in Riverside, 1900–31," *California History* 61 (Spring 1982): 27.

23   Phantasmagorias were nineteenth-century spectacles of visual wonder, using optical illusion (from magic lanterns). Frankfurt School critics of culture industries also used the term to describe the process of commodity fetishism, which makes the descriptor particularly fitting for both the Mission Inn and its acts of SCR wizardry. Theodor Adorno defined the phantasmagoria "as a consumer item in which there is no longer anything that is supposed to remind us how it came into being. It becomes a magical object, insofar as the labor stored up in it comes to seem supernatural and sacred at the very moment when it can no longer be recognized as labor." Tom Gunning, "Illusions Past and Future," http://www.mediaarthistory.org/refresh/Programmatic%20key%20texts/pdfs/Gunning.pdf (accessed January 2, 2017).

24   See Nathan Gonzalez "Riverside, Tourism, and the Indian: Frank A. Miller and the Creation of Sherman Institute," *Southern California Quarterly* 84 (Fall/Winter 2002): 193–222; Clifford E. Trafzer, Matthew Sakiestewa Gilbert, and Lorene Sisquoc, eds., *The Indian School on Magnolia Avenue: Voices and Images from Sherman Institute* (Corvallis: Oregon State University Press, 2012); and Kevin Whalen, *Native Students at Work: American Indian Labor and Sherman Institute's Outing Program, 1900–1945* (Seattle: University of Washington Press, 2016).

1

Aaron Betsky

The Wall and Its Dissolution

The Spanish Colonial Revival from Style to Vernacular

So the predominantly Anglo-Saxon culture of [Southern California] is deeply entangled with remnants of Spain.... This ancient entanglement... provides psychological support for the periodical outbursts of... the elusive but ever-present Spanish Colonial Revival style, in all its variants from the simplest stuccoed shed to fantasies of fully-fledged Neo-Churrigueresque. Such architecture should never be brushed off as mere fancy-dress; in [Southern California] it makes both ancestral and environmental sense, and much of the best modern architecture there owes much to its example.

—Reyner Banham, *Los Angeles: The Architecture of Four Ecologies* (1971)

FIGURE 1 Mature Spanish Colonial Revival residence, Castle Reagh Place and Magnolia Avenue, Riverside, built 1928. Robert Spurgeon, architect

The Wall

First comes the wall.[1] That is the essential part of the family of styles I will collectively call Spanish Colonial Revival. That wall is white, covered with stucco, and (seemingly) heavy. It keeps the outside world, whether it be natural or human-made, and which may be hostile, out. On the inside, it creates a private scene, a world all its own, shaded and sheltered so that life can flourish there. This is the interior realm proper to those who have conquered territory or can afford to live on a secluded slice of real estate they own. The SCR, above all else, marks the place of the wealthy and powerful, or those who would aspire to be so. It defends and represents what they have made for and of themselves.[2]

To enter into that world is a privilege, which the architect notes and even announces with flourishes around the doors. These portals and the few windows are the main site of decoration. To view from the interior to the outside world is also a luxury, and one that becomes the second site for decoration, as window frames dissolve into interlaced patterns. Beyond these moments of connection, a tower might signal the importance of the place, while hinting at the riches inside; at the same time, planting integrates the wall into its surroundings. Such is the basic building block of the SCR: a form that celebrates a difference between inside and outside (fig. 1).

This simple act developed into not just one way of designing and making buildings, but a tradition that now stretches back for centuries and shows no signs of disappearing. It has sprouted a number of different variations and modes of making, as well as being the discipline for the appearance of buildings from homes to office buildings, and from fast food establishments to courthouses. It has become shorthand for regressive or romantically revivalist architecture all across the Southwest, but particularly in Southern California, and is by now so diffuse and thin in its application that it is often difficult to recognize. Yet, it also still carries a coherence and a purpose that connects people to place in a particular manner.[3]

There are countless variations in building styles that have received and still receive the name,[4] but the SCR that interests me here was a style of making architecture that started in California in the last quarter of the nineteenth century, spread from there all over the world, and continues to have its adherents to this day. It was already actually several styles, each of which used some aspect of architecture that could trace its elements back to Spain or its colonies in the New World, but each of which was different in its modes of appearance and its application in the human-made environment. It was, moreover, a style with a purpose: to invent a scenario for the newly arrived middle class that provided a comfortable space—one that seemed both grander and more sensual than what they had experienced before. I am here focusing, in other words, on essentially a middle class, domestic style with clear precedents and a coherence, which is embedded within a much wider use of Spanish, Spanish Colonial, and Mediterranean motifs and elements.[5]

The nature of the scenarios the SCR developed differed over time and according to the uses to which its clients and designers put its forms. The flexibility of these components could adapt itself to the creation of civic centers, private Edens for the wealthy, or Taco Bells, serving to make each of these instances appear to be what they were not, which was part of historic and spatial continuum, as well as continuation of the middle class realm. This essay will define the basic components of the SCR style from this perspective, identify its sources, discuss some of the ways in which they combined to create structures that defined a way of living in California and other places, and show their evolution. It will speculate on its uses and on what makes it a style,[6] arguing that it has developed from an esoteric mode of appearance, to such a style, and then into a vernacular that is mass produced and anonymous. Finally, this essay will ask what, if anything, the SCR means for a designed environment that in this century must confront different social and natural conditions than what has given its reason for existence.

FIGURE 2 Santa Barbara County Courthouse, constructed in 1929. William Mooser, III, architect

The Missions

The *locus classicus* of the Spanish Colonial Style in the United States, and thus the place to which we have to turn to understand its basic components, is the Mission: the twenty-one enclosed and fortified structures that Franciscan monks, originally under the direction of Fra Junípero Serra, built, starting in 1769, up and down the California coast. These were religious communities, but also productive farms and fortresses from which the Spanish spread their culture, agriculture, religion, and political dominion over what was only to them a new place. They did not follow the local ways of making buildings, but were simplified and modified versions of similar structures in Mexico and elsewhere in the Spanish territories.[7]

Ultimately, however, the sources for the SCR are even older. The mission revived the form and function of the medieval monasteries.[8] They were religious versions of forts that served to occupy wild land perceived by the monks to be inhabited by barbarians or savages. From there, cultivation spread, both in a spiritual and a direct, agricultural sense. At the intersections of those two definitions of culture, artifacts from household implements to objects of devotion represented the embedded knowledge by which the monks did their work of colonization. They also represented their values and gave them visible shape.[9]

Although the missions developed in different ways from their first appearance in 1769 until they were secularized in 1832, and certainly became more elaborate in form and decoration, over time, the missions themselves were simple affairs in which, as noted above, the wall was the most important element.[10] Made out of adobe or a similar compound of earth and straw, they were an abstraction of the landscape rising into the vertical dimension to create a defined place. Within their confines was the world of order, comprising both gardens marked out in square plots and the cells where living and working took place.

The mission church was the most expressive element of such compounds. It connected the settlement to a larger and longer tradition, which it made evident with its orientation towards the east. Its bell tower announced the presence of a new culture, and its large interior form offered a space unlike that most of the new parishioners could experience anywhere else: cool and communal, directional and hierarchical, it turned towards the altar, whose elaborate iconography, realized in carved and painted wood, gave concrete shape to the abstractions both of religion and of the Spanish culture that carried the Catholic faith with it to the colonies.

These basic elements continue to resonate through the SCR, no matter what its forms or scale might be a century later. The perception and thus use of those pieces became colored, of course, by the fact that those employing them found them, more often as not, as ruins: the crumbled walls, eroded colors, and partial enclosures, as well as the ways in which these ruins were sometimes inhabited by different uses of a smaller scale and more heterogeneous kind, made them more malleable.[11] They had often already been elaborated into other uses. As a result, the mission was rarely replicated as a complete type, though there certainly are exceptions—its reinvention as the Santa Barbara Courthouse by William Mooser between 1919 and 1921[12] and a rambling library right next door to its original model by Michael Graves in the San Juan Capistrano in 1983[13] are both excellent examples of the possibilities inherent in the whole complex (fig. 2).

Instead, both the mission's bits and pieces and its softer, more incomplete and hybrid nature proved useful. The wall itself remained the most important part of the fragments Anglo architects reused. In contradiction to the objects sitting in the landscape that Anglo settlers brought with them beginning in the middle of the nineteenth century, the wall was a landscape element. The United States had, until then, been settled, as J. B. Jackson most clearly showed,[14] by placing individual objects in and against the landscape. By the time the settlers reached across the continent and to California, that object was a standard type: the "Stick Style" frame construction box, built with 2-×-4 timber in a modular manner.[15] It was vertical in its proportion, covered with wood, and presented façades with regular patterns of windows, a porch entry, and a gabled roof. Although the form began to relax rather quickly into the climate and landscape of California, turning into the more rambling, horizontal, and Asian-influenced bungalow (a word of Indian derivation), it retained its essential properties.[16]

The problem became clear before too long: California, at least in the coastal regions that became most heavily populated by Anglo settlers, did not have a great many hardwood forests. Moreover, the climate was not conducive to the imported style's closed and vertical forms. And, what mattered most, the structures looked out of place, while not providing the kind of space that was one of the great promises of the West Coast: the Eden of sheltered outdoor gardens and patios that could be enjoyed all year long.[17]

The responses came in the form of opening up the box to allow light and air to enter, making the bungalow into a network of redwood lattice that let inhabitants move from the darkness and shelter of the house's core out into the gardens surrounding the structure—a trend carried furthest by the Greene Brothers in and around Pasadena—but also in type. The bungalow court[18] was the most obvious innovation Southern California offered, as it broke up not the single family home, but the apartment building, using the attractiveness of the outdoor space to create a variety of spaces whose public and private nature shaded into each other while offering inhabitants more identifiable elements as their home than what was available in the standard apartment block.

This dissolution of the box in and through the bungalow and the built ruins of the walled compounds of the SCR became a mainstay of more avant-garde or experimental architecture throughout the Southwest. It resulted in the fragments of walls Rudolf Schindler poured on site and tilted up to demarcate the space of his Schindler-Case house in West Hollywood, of 1920–22 and then tied together to shelter rooms whose canvas shades opened up into private outdoor spaces for each of the house's four inhabitants. It became regularized in both Schindler's later work[19] and in that of his compatriot and fellow Frank Lloyd Wright-disciple Richard Neutra,[20] eventually turning into the steel-and-glass networks *Arts & Architecture* magazine[21] popularized during the 1950s and produced in its Case Study program.[22] In the 1980s, Frank Gehry and others broke up even that tradition to create fragments assembled not just out of the bits and pieces of stucco walls and stick construction, but also out of chain-link fence, corrugated metal, and the other materials that by then made up the landscape of Southern California.[23]

The natural response to California, both because of the climate and because this was the limit of the expansion of Anglo culture (and European culture in general), where it met and mingled with Asian and Central American traditions and people, was thus one of dissolution devolving from shade and protection, which also meant a more open relationship between the individual and the social and natural context. This fluidity could reach social extremes, but also ones in terms of architecture, turning the State into a hotbed for experimentation for several generations of architects seeking to build a more utopian society through the loose assembly of building blocks.[24]

While some clients might buy into this vision, however, the mainstream practice chose a mode for this dissolution that held onto the myth of social cohesion and differences between public and private, while erecting structures that rooted this more conservative attitude in local traditions. There was, in other words, a response that many architects realized came out of both the landscape and the relationship of the new immigrants to that place in a more organic manner, and one that was less strange or alien, because it built—at least notionally—on something that had been constructed there long before.[25] Those who chose to be at home in SCR structures, and those who designed them, were not necessarily trying to recreate Catholic monasteries (as, for instance, Ralph Adams Cram was happy to do on the East Coast),[26] as they were looking for a set of forms neutral enough, and yet specific enough to site and function, as to let them adhere to something solid and seemingly permanent.

Thus the fragmentation of the missions from a coherent type into a collection of spatial, structural, and decorative elements created a menu not unlike, and closely linked to, classicism,[27] but one that offered a sense of coming from the place, rather than being an importation. It gave the middle class who populated California a combination of belonging and propriety, openness to the new and security, and an identity with roots. This, then, was the new style: an amalgamation, as all styles are, of other elements, coming into coherence in response to a certain time and place, and to a way in which the people who used it wanted to appear.

Spanish Colonial Revival

The elements, as mentioned above, were simple, but varied. They included the stucco wall, which could be both an enclosure for a building and one for a garden or courtyard; the decorated entrance; the tower or turret; decorated window openings; tile roofs; interior spaces with wood-beamed ceilings like the mission churches; the use of colored tile with Moorish patterns to emphasize particular elements; and the overall strategy of demarcating a whole site with a wall, and then building out that enclosure into the spaces for living and working. In later times, other elements came into play so often that they became part of the canon. These included the expansion of the tower downward to become a circular staircase; crenellations on the underside of overhanging second floor walls; arched windows for major interior spaces; frequent slight changes of level; red-tile floors (which were reserved for only very few spaces in the original missions); an emphasis on the thickness of walls through the placement of windows and doors deep within their frames; and the telescoping out of spaces from the central volume in a manner similar to New England Colonial homes, but in this case often angling or even rotating around to enclose, at least partially, a courtyard.[28]

The application of these elements was most complete in single family homes, and middle class ones at that, for the simple reasons that, on the one hand, all these pieces of what came to be a recognizable style were applicable in these settings and, on the other hand, because middle class clients could afford their use. The style thus formed out of the bits and pieces of the missions (from the whole history of their existence from 1769 to 1832) developed slowly, and through admixture with a whole host of other forms. In the case of the home these included, as noted above, the Stick and Shingle Styles, but also the various forms of Colonial Revival, bungalows, and elements that had appeared in French châteaux and hotels—the staircase tower being one clear example—and were adapted to the style at hand.

What is remarkable is how coherent the style became in a short period of time. It appeared in many different places and with many different architects, but it had many of the features described above. This fast coherence is especially of note when compared to the development of the Colonial Revival, which melded many of the elements to be found in New England and English homes into a similarly more or less coherent style, but did so over the course of many decades and with much more variety, at least until the style became a post-World War II canon exemplified by the home in *Mr. Blandings Builds His Dream House* and various television shows.[29]

Certainly the SCR took some time to define itself. The revival of the missions and their use dates back to the 1890s and has its roots in the evocative paintings of Henry Sandham and others did of the missions in the early 1880s. Early nods towards the Spanish Colonial in the design of Stanford University (Shepley Bulfinch Rutan, 1887–91) and, even before that, in the Hotel Alcazar in St. Augustine (Carrere and Hastings, 1888–89), and countless softening of Beaux-Arts buildings throughout the late nineteenth century, pointed towards and drew from the missions, but the style first flourished into a confused, but evocative whole during the period of the construction of Riverside's Mission Inn, which Arthur Benton began designing in 1902 and which developed its many wings over thirty years.[30]

Architects such as Wallace Neff, John Byers, and John Kaufman can take some of the credit of developing the pieces of the style in their homes.[31] The SCR home based on a perhaps vaguely understood mission precedent really came into coherence when Santa Barbara adapted a streamlined version of it by civic mandate after the 1925 earthquake leveled much of the town. Certainly the work of George Washington Smith in and around Montecito became the style's most sophisticated elaboration, offering, as David Gebhard has convincingly argued,[32] the most thorough integration of its components, while melding them into compositions that responded to site and program with a clarity that made them seem like natural, if not wholly organic, California homes. Yet, already in the decade before these rather grand designs appeared, home owners across the Los Angeles, Riverside, and San

Diego areas had been able to buy both developer homes and architect-designed SCR houses that integrated much of the style's characteristics. The style had even trickled down to mass working housing created by the railroads for their workers.

One of the most perfect examples of the spread and variety of the domestic version of the SCR style is still on view in Riverside's Wood Streets neighborhood.[33] There the architects Robert Spurgeon and Henry L. A. Jekel designed dozens of homes in the 1920s that line the flat, otherwise featureless subdivision. Towering palms stand above, presenting countless and sophisticated variations of the walls, windows, turrets, and other elements described above, each fit into the small lot proper to the middle class budget. There are no "pure" examples among the lot, as both architects came from other places and backgrounds, and put some Arts and Crafts, some Tudor, and Beaux-Arts pieces into the SCR mix. The result, however, is that the compositions do not seem to confine themselves to single homes, though they do resolutely define each middle class terrain, but rather echo and complete each across the street and along the block, creating a sense that this a neighborhood with coherence and clarity that matched what you could find in those East Coast and Midwest neighborhoods that have been developed with various mixtures of Colonial Revival styles in previous decades. This was the new middle class terrain, conquering the landscape with as much confidence and clarity as the missions had, though now in many different pieces.[34]

Like the Colonial Revival, much of the SCR's popularity after the late 1920s came from the influence of popular media, especially travel and shelter magazines such as *Sunset*, and from movies. Films such as *The Mark of Zorro* (1920) brought back the romance of "Old California," as had books such as Helen Hunt Jackson's 1884 *Ramona* in a previous generation. Wealthy film stars were portrayed living in SCR homes, and in films such as *Double Indemnity* (1944) the middle and upper middle class almost invariably live in such structures.[35]

By the 1920s, the SCR had thus become a coherent domestic style that rivaled the Colonial Revival in popularity in Southern California, and far outstripped such other contenders as Tudor, Craftsman, or French Provincial, though each of those had pockets of popularity and became admixed with the two dominant styles. While the Colonial Revival had spread west, the SCR moved in the reverse direction, eventually showing up in suburbia all over the country and developing, under the hands of architects such as Addison Mizner,[36] local variations appropriate to the climate, conditions, and traditions of places such as Florida.

What is especially interesting is what the style said about its middle class inhabitants. It gave them homes based on the heritage not of their own ethnic forebears (as there are very few middle class families of Hispanic heritage; most of the vastly enlarged Hispanic population of the immediate post-war period having come in as agricultural workers and economic immigrants),[37] but of that of a culture that was essentially alien to their history and tradition. Though you could argue that, say, an Irish family was just as transplanted in a home built up out of English elements, the distance from the source was incrementally larger in the SCR. It marked a further liberation of American culture from its mainly northern European roots. One might even wonder (and this is pure speculation) whether the very alien nature of the forms helped make people from different backgrounds, such as Jews or Eastern Europeans, feel more at home in this environment.

After all, the Colonial Revival and Tudor styles so prevalent on the East Coast had a strong association with one country, England, whereas the SCR less specifically referenced on place and, to the extent that it did, it was a place that was as alien to mainstream Anglophone American culture as many of these new immigrants may have felt themselves.[38]

Through the presence of elements that came out of the workshop traditions of the missions, but were reinforced by the strong influence of the Arts and Crafts movement in the SCR,[39] the homes also became places that brought craft into the daily life of the inhabitants. Tiles with patterns that went back to Moorish, Turkish, and Iranian roots, as well as wood furnishings and window surrounds that were carved with floral patterns or deeply incised, and that were assembled with visible peg connections, gave the home the sense that it was not fully mass produced, but was part of a realm of culture, reinforcing its role in middle class culture as a place of acculturation and learning through "artifactual literacy."[40] Finally, the roughness of textures, such as the stucco walls (sometimes even with brick or stone deliberately showing through in places), the tile floors, and wood details, gave the homes a sensuality that rooted them in place, further removing them from the alienating world of work, shopping, and play that was the life of most of the homes' inhabitants. The gardens, whose lushness continued the myth of California as a New Eden that had been promulgated by railroad companies and real estate agents in their push to sell the state, further reinforced all these sensibilities and messages.[41] The fact that the SCR home was better at embracing and even enclosing the garden—again a formal attribute central to its mission roots—made its presence both more natural and a more integrated part of the house.

Over the decades, the SCR shed more and more of the specific elements that gave it the particular character and power described above, and, by the time that it had turned into the "Red Tile Tide" that covered so much of California starting in the 1970s, only that roof, the use of stucco (or later substitutes such as Dryvit), and a few ornamental details distinguished it from the many other weakened strands, inoculated for selling and standardized for production, that made up the hodgepodge of homes developers plunked down on the landscape with little concern for their sites. Only the very wealthy could afford something more like the original type, and commissioned periodic revivals of the style that showed its versatility and sensuality to good effect. The works of several of the contributors to the present volume are good examples of the latest such version of this by now well-worn and well-ornamented style.[42]

Civic Forms

Though homes form the SCR's mainstay, we should not forget about the style's civic variations. The fact that the style there has deteriorated to enlarged versions of Taco Bells that adorn quite a few suburban libraries and community centers or town halls, not to mention countless speculative office developments, should not detract us from their original power to define and fix middle class values in California any more than the devolution of the mission-derived homes should.

The civic variant of the SCR helped define a new form of cohesion for governmental and institutional buildings that was more romantic and more open than the palaces, of a French and ultimately Roman derivation, in which such structures were housed in most of the country and which still remained popular in California throughout the period until World

War II. As noted above, the first step in this movement was the dissolution of civic structures into more literally open complexes that invited in the outside air, enclosed courtyards, and created places of gathering that were less defined and demarcated that they might be in Beaux-Arts style buildings.

Over time, even public buildings began to have this sense of openness. Though the Pasadena City Hall, for instance, is a symmetrical building whose major elements are disposed on its full-block site according to Beaux-Arts principles of hierarchy and composition, most of the site actually consists of open courtyards. Even the entrance portico, whose dome and grand arch makes it appears to be a monument to civic power, is a shaded, but outdoor, space.[43]

Civic buildings in the SCR style, however, looked not only to the missions, but also to the stage set-like structures Bertram Goodhue designed for the 1915 Panama-California Exposition in San Diego's Balboa Park.[44] Together with the Panama-Pacific Exposition of the same year in San Francisco,[45] master planned not by an architect, but by the painter Jules Guerin, these fair structures, which survived today at least in part as museums, park structures, and bridges, the Panama-California Exposition undressed classical forms into something much more primitive. Goodhue reserved ornament for the building's edges, treating them as abstract objects that melded into the existing cliffs and rose up into a floriated dissolution against the sky.[46] The fact that these buildings were not meant to be permanent structures made them looser and more open, while their responsiveness to the landscape as objects stood in contrast to the enclosing forms coming out of the missions.

The most extreme example of this tendency, even more open than the Pasadena City Hall, is the San Francisco Exposition's only remaining structure, the dome of the Palace of Fine Arts. Designed by Bernard Maybeck, its central element is a classical dome in front of an abstract, curving volume. This functionless object is a gigantic folly, serving only as a marker of civic significance. Underneath it the park and its small lake flow around columns and bases that give the structure a sense of being either a ruin or the building block for a future California civilization.[47]

Everywhere around California you can thus find grandeur softened with and into the landscape, and dissolved into something more tentative and open-ended, but also more nostalgic, than the French-based institutional buildings that serve as civic centers in so much of the rest of the country. Yet the public version of the SCR not only made power seem more natural (in every sense of the word) and less alien to those who were cultivated and empowered to use them, it also made its forms seem more familiar.

When Mooser designed the Santa Barbara Courthouse, however, he also gave it a character that helped the middle class feel as if their derived and collective power had, like their middle class domestic lives, deeper roots than their new arrival in the area would seem to warrant. Part of the force of this and many other SCR civic structures derives, I believe, from its domestic character. Instead of ranks of columns and pediments forestalling entrance and making the institutions housed behind them seem grander and of a completely different material, quite literally, than what its users might experience every day, the Santa Barbara Courthouse spreads a large arch to welcome inhabitants, and then rambles around two corners of what is otherwise an open and public park with a succession of court rooms and offices. Its most prominent roof signals a church-like space, which that is indeed what the major public gathering space resembles, while the bell tower proclaims its civic function. Beyond such markers, however, most of the other elements are of a much more familiar kind and scale: a protruding turret that contains a circular staircase, windows surrounded with tile, and telescoping forms all should make Southern California visitors to the Courthouse feel at home. The Courthouse is an expansion of the middle class' realm into the civic, not an alternative to their retreats.[48]

Few other civic structures in California matched the Santa Barbara Courthouse in transforming a civic structure so thoroughly into another type, although there was a revival of this approach to government buildings starting in the late 1980s. The embrace of public green spaces, the opening up of monolithic structures (though limited by the need for security and the prevalence of conditioned air), and the application of certain crafted details (more sparse and often mass produced) marked especially the design of Moore Ruble Yudell's 1982–90 Beverly Hills Civic Center, but is also evident, for instance, in Escondido's Art Center (1988–94), by the same architects.[49]

From Style to Vernacular

These days, it is difficult to distinguish and evaluate both these distinguished examples of the Spanish Colonial style at civic level and the fine examples of its domestic variants, because the style has become a vernacular.[50] It is now not a conscious choice for architects, clients, or buildings, but just one of the ways you build in Southern California (and elsewhere in the country and even world). Its forms have become standardized, and therefore cheap to make. Its elements are also mass produced and easy to find at the local Home Depot. Because the forms are so much part of the background landscape that make up the millions of square miles of urban sprawl into which much of the state has turned, we do not even notice the bits and pieces themselves.

What is even more important, the power the SCR once had to make its inhabitants feel at home in this particular new-found land has disappeared in much the same way in which the Colonial Revival lost that ability on the East Coast. The style is hardly even a marker of middle class achievement, either, as its signals and clothes mark strip malls and fast food establishments, as well as acres of very cheap housing. It has, instead, become the stuff in which millions of people live, work, and play. Only occasionally, as in the Mondavi Winery's (designed by Cliff May, master of the ranch house, in 1964, and completed in 1966) evocation of the Santa Barbara gateway, or in the appearance of grand SCR homes as sets for movies and television programs, do we see a more contemporary expression of the distinct style as a sign of place, as well as of middle class power, accommodation, and accumulation of wealth. If a style lets you define yourself, a vernacular makes you disappear, and that is what the SCR does today: form the anonymous backdrop to anonymous lives.

The very rich can still afford the style with all its accouterments, in the manner that they can afford couture clothing, and may choose finely made traditional forms over avant-garde expressions of their taste and distinction. The SCR today, however, is a dead style, one that barely permits any adaptation to current technologies and different modes of life, however beautifully its structures have been made. It is only historical examples from the period when it was still a style that make this mode of building of interest to us today.

We could try to speculate on ways in which the style could be revived as a manner in which we could create homes and other structures that are more integrated into the landscape, crafted in a way that might connect us to the making of things in a world of technological dominance, and molded to define a new middle class.[51] A built argument of this sort was made at the largest scale and in the most sophisticated manner I know recently in the development of the Playa Vista area of Los Angeles, and yet the results, despite good intentions and the work of some good designers, are depressing: cookie cutter blocks crammed together, lightly covered with SCR markers, and devoid of any of the sense of space to which the style aspired—let alone of the kind of lush semi-private and public open spaces that are central to its success.[52]

The problem is that the economics of site use and production of buildings make this almost impossible,[53] and the California landscape is, because of global climate change, becoming a very different place.[54] It is also true that the middle class as it existed in the period between the middle of the nineteenth and the middle of the twentieth century has dissolved into something much more amorphous, mobile, and diverse.[55] It is a class that defies definition and it would be difficult to imagine a style appropriate to hipsters, aspiring working class immigrants from Honduras or China, hardworking factory workers, office drones, and spoiled millennials that would make sense in the way the SCR did for a large group of Anglo immigrants living and working in Southern California a century ago.

For all these reasons, it would be better to admire these missions and remnants of fairs, these homes small and grand, and these crafted elements and integrated gardens for what they are: a reminder of the romance of California as a middle class paradise for Anglo settlers.

Notes

1. Although there is no direct connection to the architecture of which this article speaks, I am indebted for this phrase to William Morgan's *The Almighty Wall: The Architecture of Henry Vaughan* (Cambridge, MA: The MIT Press, 1982). The relationship between neo-Gothic and Spanish Revival Architecture, however, will be discussed below.
2. There is, to my knowledge, no thorough architecture analysis of the California Missions. There are early articles such as Rexford Newcomb's "Architecture of the California Missions" [*Annual Publication of the Historical Society of Southern California* 9, no. 3 (1914): 225–35] often cited, and chapters in many of the books on California architecture, but I base much of the following on my own observations.
3. For a broad survey of SCR architecture, see S. F. Cook and Tina Skinner, *Spanish Colonial Architecture* (Atglen, PA: Schiffer Publishing, 2005). For its larger cultural context, see Kevin Starr, *Americans and the California Dream 1850–1915* (New York: Oxford University Press, 1973), 365–414. For a broader cultural background, see Kevin Starr, *Inventing the Dream: California through the Progressive Era* (New York: Oxford University Press, 1986), 3–98.
4. The definition of SCR is a difficult one. I use it here in the most generic sense possible, as part of an agreement for a common nomenclature for this book project. The source for the definition remains David Gebhard, "The Spanish Colonial Revival in Southern California," *Journal of the Society of Architectural Historians* 26, no. 2 (May 1967): 131–47. See also Karen J. Weitze, *California's Mission Revival* (Santa Monica: Hennessy & Ingalls, 1984).

5   Cf. Starr, *Americans; Inventing the Dream*, loc. cit. This discussion is for me also part of a larger one about the manner in which middle class designers and clients appropriated, started in the middle of the nineteenth century, historical styles in such a manner as to make a place for themselves in the modern world. See my *Making It Modern: The History of Modernism in Architecture and Design* (Barcelona: ACTAR, 2015).

6   The question of style in architecture is obviously a complex one, first raised explicitly as a central variable in architecture (as opposed to an extension of an adapted canon or mode of design, or a fashion) at length by Gottfried Semper in his never-completed treatise of that name of 1860–62, recently translated as *Style in the Technical and Tectonic Arts; or, Practical Aesthetics*, trans. Harry Mallgrave and Michael Robinson (Los Angeles: The Getty Research Institute, 2004). The notion that the appearance of a building is both independent from its formal and structural properties, and is a question of choice, has since then troubled architects seeking to justify their work. The notion of a coherent style with both an economic and social justification and a developed canon or coherent accumulation of elements will define the word for the purpose of this article.

7   For an early survey of Spanish Colonial architecture that was of some influence in the development of the revival, see Sylvester Baxter, *Spanish Colonial Architecture in Mexico* (1901), of which one volume was reprinted in 2015 by Forgotten Books; several other volumes are also available as reprints. See also Dora P. Crouch, Daniel J. Garr, and Axel I. Mundigo, *Spanish City Planning in North America* (Cambridge, MA: The MIT Press, 1982).

8   For the architecture from which the standard Benedictine, as well as later other, monasteries developed, see Walter Born and Ernest Horn, *The Plan of St. Gall: A Study of the Architecture & Economy & Life in a Paradigmatic Carolingian Monastery* (Berkeley: The University of California Press, 1979).

9   For an early description of the Missions, see Henry Miller, *Account of a Tour of California Missions & Towns* (Los Angeles: Bellerophon Books, n.d. [1856]), as well as Rexford Newcomb's more popular *Spanish-Colonial Architecture in the United States* (New York: Dover Publications, 2002, 1937).

10 For a survey of how the Missions developed during this period, see Edna Kimbro, Julia Costello, and Tevvy Ball, *The California Missions: History, Art and Preservation* (Los Angeles: Getty Conservation Institute, 2009).

11 For recent discussions of the historical uses of ruins, see Christopher Woodward, *In Ruins: A Journey Through History, Art, and Literature* (New York: Vintage Press, 2010); and Thomas J. McCormick, *Ruins as Architecture: Architecture as Ruins* (St. Petersborough, NH: Bauhan, 2000).

12 Cf. Patricia Gebhard and Kathryn Masson, *The Santa Barbara County Courthouse* (Santa Barbara: Daniel & Daniel, Publishers, 2001).

13 Cf. Karen Vogel Wheeler et al., *Michael Graves: Buildings and Projects 1966–1981* (New York: Rizzoli International Publications, 1982).

14 John Brickerhoff Jackson, *Discovering the Vernacular Landscape* (New Haven: Yale University Press, 1984); and *A Sense of Time, A Sense of Place* (New Haven: Yale University Press, 1996).

15 Vincent Scully, Jr., *The Stick Style and the Shingle Style: Architectural Theory and Design from Downing to Wright* (Gloucester, MA: Peter Smith Publisher, 1971).

16 The most complete history of the bungalow's origins and spread in California remains Robert Winter, *California Bungalow* (Santa Monica: Hennessy & Ingalls, 1980).

17 Cf. Starr, *Inventing the Dream*, op. cit., 128–75.

18 Cf. Stefanos Polyzoides et al., *Courtyard Housing in Los Angeles* (New York: Princeton Architecture Press, 1996).

19 David Gebhard, *Schindler* (Layton: Peregrine Smith, 1971) offered the first monographic review of his work, while Judith Sheine [*R.M. Schindler* (London: Phaidon Press, 2001)] offers the most complete. See also Robert Sweeney, *Schindler, Kings Road, and Southern California Modernism* (Berkeley: The University of California Press, 2012).

20 Thomas S. Hines, *Richard Neutra and the Search for Modern Architecture* (Berkeley: The University of California Press, 1982).

21 The complete run of the magazine was reprinted by Benedikt Taschen (Cologne) in 2014.

22 Esther McCoy, *Case Study Houses 1945–1962* (Santa Monica: Hennessy & Ingalls, 1977 [1962]); see also her *The Second Generation,* (Salt Lake City: Gibbs Smith, 1984).

23 Peter Arnell and Ted Bickford, eds. *Frank Gehry: Buildings and Projects* (New York: Rizzoli International Publications, 1985).

24 This tradition of experimentation started with Frank Lloyd Wright's textile block houses, continued with the work of Neutra and Schindler, who moved to California to work on those projects, and then through the modernist Case Study movement described above. It welled up again in and around the Southern California Institute of Architecture and the work of Frank Gehry starting in the late 1970s, and shows no signs of abating. See Esther McCoy, *Second Generation*, op. cit., and my own *Experimental Architecture in Los Angeles* (New York: Rizzoli International Publications, 1992).

25 This is the argument Michael Burch makes in his unpublished manuscript "A Return to Authenticity: The Spanish Colonial Revival at 100."

26 The most complete, if not uncontroversial, discussion of Cram's work and theories can be found in Douglass Shand-Tucci's *Ralph Adams Cram: Life and Architecture* (New York: W. W. Norton, 1995); and *Ralph Adams Cram: An Architect's Four Quests Medieval Modernist American Ecumenical* (Boston: University of Massachusetts Press, 2005).

27 In many ways, the SCR is a sub-style of classicism. Its use of formal elements derived from the classical canon is extensive, and includes all those mentioned above, but also columns, pediments, and friezes. I here would argue that the coherence of the variations it developed, as well as the admixture of vernacular elements and the use of such building techniques as adobe (or imitations thereof) warrant classifying it as a style into itself.

28 For a useful survey of the style, its sources, and its permutations, see Arrol Gellner and Douglas Keister, *Red Tile Style* (New York: Viking Studio, 2002).

29 Vincent Scully (see above) first discussed the use of various revival styles in his *The Shingle Style Today: Or The Historian's Revenge* (New York: George Braziller, 1974). For the deeper cultural background of the revival, which helped the American middle class define themselves by claiming older roots, see T. J. Jackson Lears, *No Place of Grace: Antimodernism and the Transformation of American Culture, 1880–1920* (Chicago: The University of Chicago Press, 1981).

30 Weitze, *California's Mission Revival*, op. cit.; Maurice Hodgen, *Master of the Mission Inn: Frank A. Miller, A Life* (North Charleston: Ashburton Publishing, 2014).

31 The one monograph of this selection of architects is Alson Clark, *Wallace Neff: Architect of California's Golden Age* (Santa Monica: Hennessy & Ingalls, 2000).

32 David Gebhard, *George Washington Smith 1876–1930* (Santa Barbara: The Art Gallery of the University of California at Santa Barbara, 1964); see also Patricia Gebhard, ed., *George Washington Smith: Architect of the Spanish Colonial Revival* (Salt Lake City: Gibbs Smith, 2005).

33 With many thanks to H. Vincent Moses for showing me this neighborhood and providing essential background information.

34 This is what Reyner Banham described as the "Plains of Id" in his *Los Angeles: The Architecture of Four Ecologies* (Berkeley: The University of California Press, 2001 [1971]), 143–60.

35 Helen Hunt Jackson, *Ramona: A Story* (New York, Penguin Signet Classics, 1984 [1884]), for the best discussion of the uses of SCR homes in film, see *Los Angeles Plays Itself*, directed by Thom Anderson (2003).

36 Donald W. Curl, *Mizner's Florida: American Resort Architecture* (Cambridge, MA: The MIT Press, 1987).

37 The Bracero program in particular brought in a large influx of Hispanic workers, but the economic boom also attracted immigrants beyond such sanctioned programs. See http://ic.galegroup.com/ic/uhic/ReferenceDetailsPage/DocumentToolsPortletWindow?displayGroupName=Reference&u=oldt1017&u=oldt1017&jsid=347c2091a080fbd71aebaad95d411063&p=UHIC%3AWHIC&action=2&catId=&documentId=GALE%7CBT2313026907&zid=dbaf8355e54c396b9af1f64d3a9cea8c.

38 For an interesting alternative to such "buying in," see Del Upton's tracing of variations in American housing stock based on ethnic heritage: *America's Architectural Roots: Ethnic Groups that Built America* (New York: John Wiley & Sons, 1986). For changing attitudes towards being at home in the United states, see Lears, *No Place of Grace*, op. cit.

39 As the example of the Mission Inn in Riverside shows, the Arts and Crafts Movement was essential to the development of the SCR. Just as the Colonial Revival on the East Coast made use of abstracted versions of sixteenth and seventeenth century furniture, so the Mission Revival looked back to surviving pieces from Missions and Ranchos in California. Cf. Robert Winter, *Toward a Simpler Way of Life: The Arts & Crafts Architects of California* (Berkeley: The University of California Press, 1997), esp. 181–208, and Kenneth R. Trapp, ed., *The Arts and Crafts Movement in California: Living the Good Life* (New York: Abbeville Press, 1993).

40 This phrase was coined by Harvey Green in his lectures; see also his *The Light of the Home: An Intimate View of the Lives of Women in Victorian America* (New York: Pantheon Books, 1983).

41 See Starr, *Americans*, op. cit. The archives of *Sunset Magazine*, portions of which were made available to this author during the course of research by the editors, offer countless examples of such garden architecture over more than a century of publication. See also Richard J. Orsi, *Sunset Limited: The Southern Pacific Railroad and the Development of the American West, 1850–1930* (Berkeley: The University of California Press, 2005), 130–68.

42 Cf. http://www.michaelburcharchitects.com/; http://www.mparchitects.com/site/landing?au=a; http://www.moorerubleyudell.com/.

43 Although Arthur Brown was a prominent architect, trained at the École des Beaux-Arts in Paris, and responsible for quite a few civic buildings in California, there are no published monographs of his work.

44 The original pictorial survey of the Exposition, *The Architecture and Gardens of the San Diego Exposition: A Pictorial Survey of the Aesthetic Features of the Panama-California International Exposition* was digitally republished in 2012 by Ulan Press. See also Richard Oliver, *Bertram Goodhue* (Cambridge, MA: The MIT Press, 1983); and Romy Wyllie, *Bertram Goodhue: His Life and Residential Architecture* (New York: W. W. Norton & Company, 2007), 118–29.

45 Cf. Laura A. Ackley, *San Francisco's Jewel City: The Panama-Pacific International Exposition of 1915* (Berkeley: Heyday Press, 2014).

46 *San Diego Exposition*, op. cit.

47 Bernard Maybeck, *Palace of Fine Arts and Lagoon: Panama-Pacific International Exposition, 1915* (Ann Arbor: University of Michigan Reprint, n.d.).

48 *Santa Barbara County Courthouse*, op. cit.

49 Charles Willard Moore et.al., *Moore Ruble Yudell* (London: St. Martin's Press, 1993).

50 Jackson, *Discovering the Vernacular Landscape*, op. cit. made the distinction in his lectures between original vernacular structures and those to which classical or other "high" elements were attached: only something made from the land and locally available materials, such as a sod house, should be considered vernacular architecture. Conversely, I would argue that a style can dissolve into vernacular when such applied elements are what is available in the hardware store and are mass produced, turning specific specifiers into clichés and standardized markers or logos.

51 This has been the argument, at least in part, of the architects associated with the Congress of New Urbanism, though they would claim that their work actually is classless; see Andreas Duany, Elizabeth Plater-Zyberk, and Jeff Speck, *Suburban Nation: The Rise of Sprawl and the Decline of the American Dream* (Boston: North Point Press, 2000).

52 http://www.laweekly.com/news/playa-vista-was-going-to-be-a-utopian-planned-community-but-capitalisms-harsh-reality-keeps-creeping-in-4378497.

53 The economic constraints on architecture are becoming more and more confining. The rise of "value engineering" and other forms of pseudo-scientific evaluations of the "worth" of architectural elements, the spread of standardized building components, and the use of computer programs designed to maximize ease of construction (BIM) make it more and more difficult to find room or funds for any kind of design specific to a site or situation, or to allow for any kind of "eccentricity" in construction. Added to this are the constraints of code, which restrict anything that might either harm or offend.

54 It is interesting to note that none of the discussions of climate change have looked seriously at its impact on architecture. It is clear, however, that persistent droughts will make it difficult for the lush gardens central to the SCR, while the closed nature of most of these designs make natural ventilation difficult. A return to the Missions (once again) might provide answers in their sparse irrigation of their gardens of native species and their use of adobe for thermal mass.

55 The dissipation and possible disappearance of the middle class since its culture was designed by Max Weber in his "The Protestant Ethic and 'Spirit' of Capitalism" in 1905 (republished New York: Penguin Classics, ed. by Peter Baehr and Gordon C. Wells, 2012) has been often noted; for the most recent coherent diatribe on the subject, see Arianna Huffington, *Third World America: How Our Politicians are Abandoning the Middle Class and Betraying the American Dream* (New York: Crown Press, 2010).

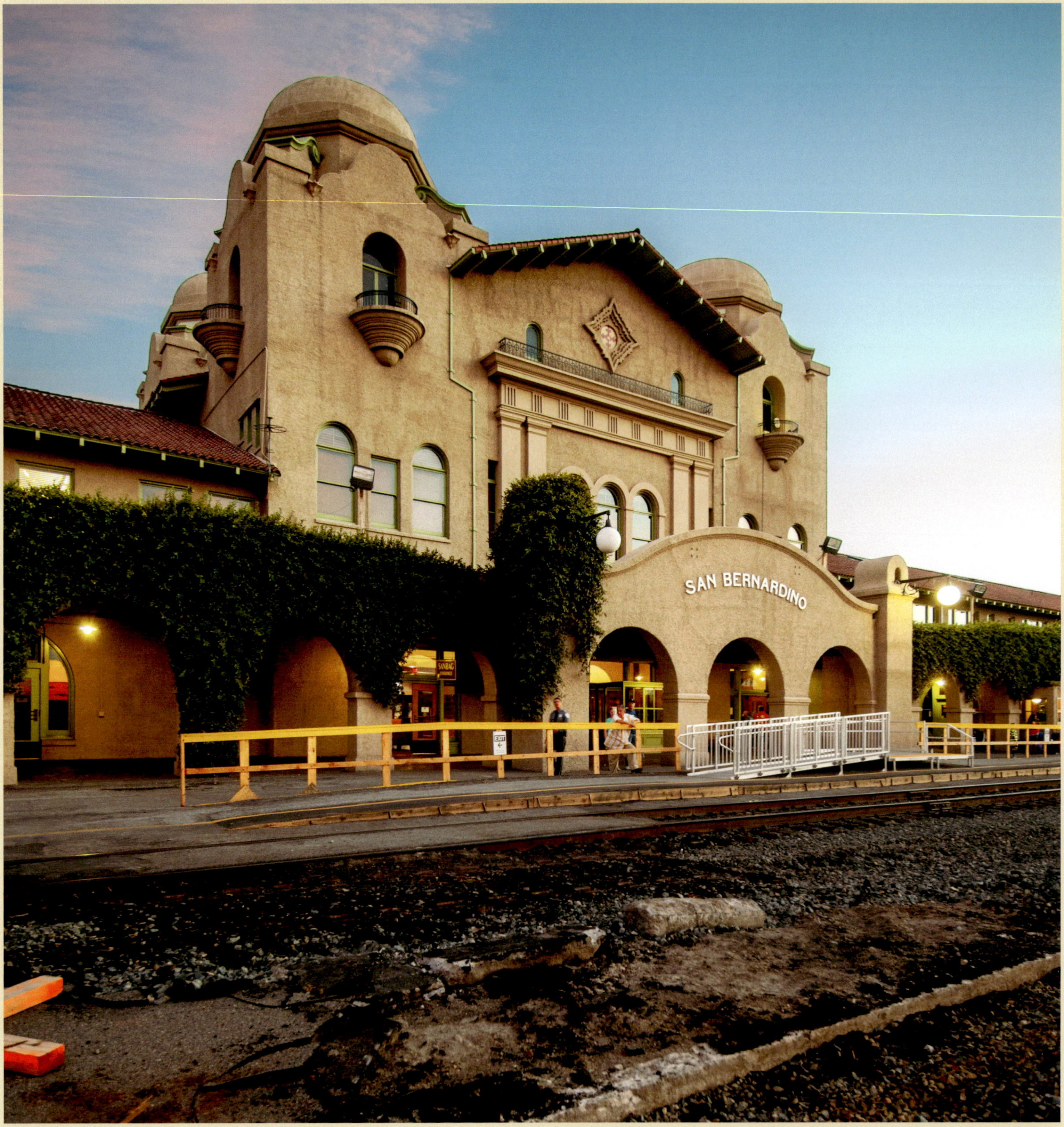
SAN BERNARDINO

2

H. Vincent Moses and Catherine Whitmore

Castillos, Iglesias y Casas

Constructing the Spanish Colonial Revival in the Inland Empire, 1895–1935

The Spanish Colonial type of architecture is that for which Riverside is ... celebrated ...

—*Riverside Daily Press*, January 26, 1927 (Building and Realty Page)

The Great Spanish Awakening in the Inland Empire

It began in 1884 as pure fiction, a tragic love story of a mythical heroine named Ramona. Helen Hunt Jackson told her story in the best-selling novel of the same name, weaving the heroine's woebegone tale neatly around a fanciful romantic view of California's Spanish mission past. Jackson intended her novel to redress perceived grievances perpetrated on Southern California's Native Americans by Protestant Anglo American interlopers, and the loss of land and status to these intruders by the area's native born Mexican Californio families. Jackson's Ramona myth quickly gained momentum, reviving interest among the Anglo newcomers in California's Spanish Colonial heritage. To Jackson's dismay, however, Ramona's fictional world emerged by the late 1890s as a powerful Anglo promotional device, used by the *Los Angeles Times*, the transcontinental railroads, Charles Fletcher Lummis, and Riverside's Frank A. Miller, to conjure up an exotic, fantasy identity for this new regional capital of Anglo population and power. The wholesale remake of Ramona by promoters quickly gave rise to what former State Librarian Kevin Starr called a "Mission Cult," built around preservation of the deteriorating California missions.[1]

By the mid-1920s, this Spanish Colonial Revival had completely conquered Inland

Southern California, the Inland Empire. In its wake, Revival-inspired building design became the primary architecture of choice in the principal Inland Empire cities of Riverside, Corona, San Bernardino, and Redlands. While traditionally identified with Santa Barbara and other Southern California coastal cities, the Revival permeated all facets of local inland society. This essay examines the complicated origins, rise, theoretical foundations, and key architectural expressions of the Revival in these cities of Western Riverside and San Bernardino Counties. For this analysis, the Inland Empire is defined by these cities. The role of the Revival in each of them is, in turn, illustrated by key iconic structures in each city, rising to the level of the National Register of Historic Places.

First, a definition: By 1925, the mature SCR, named the "California Style" by boosters and architects, was readily identified by its characteristic red clay barrel-tile roof, plaster and stucco cladding, recessed arched windows/doors, arcades, limited wall openings, large focal windows (often of art glass), decorative iron work, carved stonework, courtyards, glazed tile decoration, decorative Spanish chimney pots, domes, round or square towers, and often Monterey balconies.[2] What an architecture it was, too: broadly eclectic, inventive, laced through and through with the inner muse of the Arts and Crafts Movement, and buoyed by a mythical reinterpretation of Greco-Roman, Moorish, Spanish Mission, Spanish Andalusian Vernacular, and related Mediterranean styles.[3]

Enabled by Riverside's early twentieth-century navel orange economic boom, the Revival transformed these four Anglo-dominated citrus belt cities from typical Victorian Vernacular towns into a pseudo-Spanish "Magic Kingdom"; a Mediterranean Arcadia, awash with references to Spanish dons, missions, and Iberian castles (fig. 1). The eclectic Spanish-inspired architecture, and cultural myth making around the Revival, produced a vision of an exotic, better Mediterranean just inland from the shores of the Pacific.[4] The result was that by 1925, the area's Protestant Anglo elite had co-opted the SCR lock, stock, and barrel as its own. The transformation was so complete that it enabled Anglo adopters, such as Arts and Crafts Movement aficionado Frank Miller, to secularize Catholic iconography and building forms, detach them from their church affiliation, and put them to use in their reinterpreted neo-Spanish building program. This wholesale appropriation was rather remarkable given the intense Anglo Protestant animus toward "Papism," and the Catholic Church (fig. 2).[5] Ironically, the Ramona story had given the ruling Anglo elite a protestant

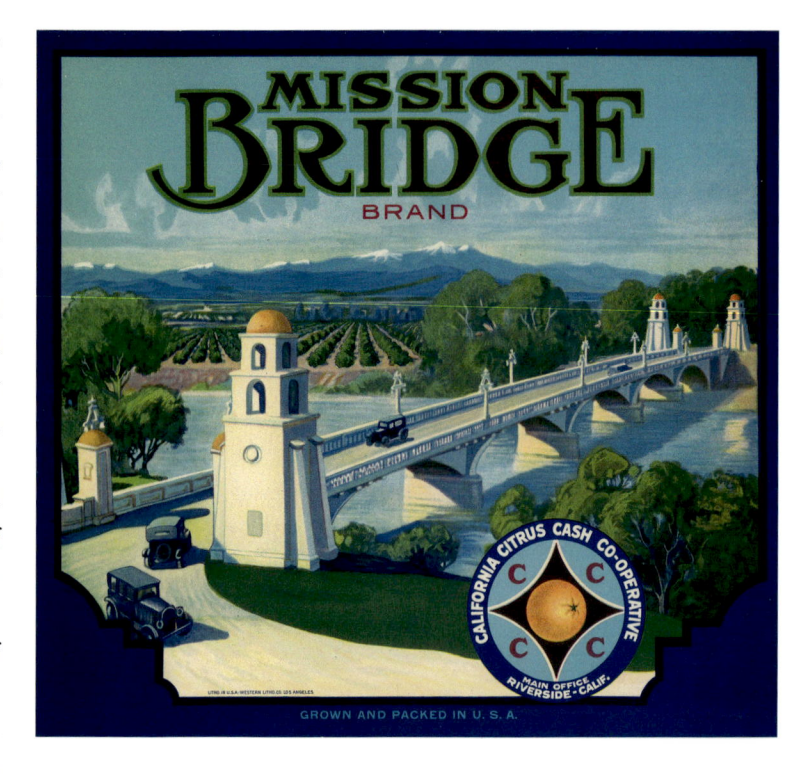

FIGURE 1   Mission Bridge Brand orange box label, California Cash Co-Operative, Riverside, depicting newly expanded 1928 Santa Ana River bridge at the western entrance to the city, ca. 1932

FIGURE 2 Theodore Roosevelt, with Frank A. Miller and Isabella Hardenberg Miller, replanting one of the parent navel orange trees at "The Old Adobe," New Glenwood Hotel, California's Mission Inn Grand Opening, May 8, 1903

morality play, which had allowed them to absorb the Native American and Spanish heritage of California as their own.[6]

In turn, the stature of the Inland Empire and its prevailing architecture grew immensely in the 1920s. California Fruit Growers Exchange (Sunkist), the railroads, and Frank A. Miller of Riverside's Glenwood Hotel (now known as the Mission Inn) seized the promotional value of the exotic "Mediterranean" imagery of the region to sell citrus fruit, and bring masses of eastern and foreign visitors and settlers to the inland area. By the Roaring '20s, Riverside and her neighboring cities were in an Anglo upper and middle class building boom to accommodate the rising in-migration.[7] The boom featured houses and other buildings in the SCR style, including upscale housing tracts, commercial buildings, and public structures, including schools, churches, and hospitals.

Origins, Rise, and Architectural Expression: Spanish Colonial Revival in Two Phases, Mission and Mediterranean Revival

The SCR in Southern California occurred in two distinct, yet overlapping phases. The Mission Revival, Phase one, took hold by 1895 and crested by 1915. Phase two, the Mediterranean Revival, arrived in earnest in the immediate aftermath of the 1915 Panama-California International Exposition, Balboa Park, which featured a bombastic form of Spanish Baroque style that rapidly swept through California.[8]

Mission Revival, 1895–1915

The Mission Revival itself began in 1895, when Charles Fletcher Lummis founded the Landmarks Club of California, an organization dedicated to the restoration and promotion of the crumbling California missions, the State's own "ruins." Lummis argued that "the Missions are, next to our climate and its consequences, the best capital Southern California has."[9] Perfectly suited to found the Club, the one-time city librarian extraordinaire who left the Los Angeles Public Library with a fabulous local history collection, had served as city editor of the *Los Angeles Times*. Later, he founded the noteworthy Southwest Museum, dedicated to the preservation of regional Native American artifacts and culture as part of his work on preservation of the missions. Lummis, who gave himself the moniker "Don Carlos," fought a tactical battle to give the region a brand, and thought the missions were key to his effort. His chief megaphone for that effort was as editor of a high-brow booster magazine *Land of Sunshine*, later renamed *Out West*, which persistently promoted Southern California as a bourgeois Hispanicized utopia—an Arcadia of unlimited opportunities just waiting for adventurous individuals to explore.

Among the founding members of the Landmarks Club were two architects: Arthur B. Benton and Sumner Hunt, both of whom made significant contributions to the adoption of "Mission" architecture as the most appropriate style for the region.[10] Benton argued in 1896 that "Mission style architecture possesses breadth and massiveness unusual in any style and much of its detail is admirably designed and executed for beauty and durability." In his view, and Lummis', mission architecture clearly suited the region's semi-arid climate and Mediterranean-like terrain, and reflected its Spanish heritage more effectively than eastern residential styles. Local promoters Harrison Grey Otis, owner and publisher of the *Los Angeles Times*; Henry Huntington, megacapitalist and owner of the Pacific Electric Railway; and Frank A. Miller of the Mission Inn then took the theme and perfected it in a persistent and intensive propaganda campaign that carried architects with it.[11]

Ironically, "Neither the essential forms nor the structure of the Mission Revival buildings had anything to do with their supposed prototypes," which were the missions themselves, argued David Gebhard. "Instead . . . architects," such as Benton with Miller's Glenwood Mission Inn in Riverside, "conjured up the vision of the Mission by relying on a few suggestive details: simple arcades; parapeted, scalloped gable ends (often with a quatrefoil window); tiled roofs; bell towers . . . and finally . . . broad, unbroken exterior surfaces of rough cement stucco." The Mission Revival architects frequently borrowed ornamental detail from the Moorish traditions of North Africa and Spain, the Richardsonian Romanesque, and the Arts and Crafts-driven design detail of the great Louis Sullivan, and the "Mission" furniture and Craftsman interiors of Gustav Stickley. In other words, Mission Revival was an almost entirely made up architectural style. The eclectic fantasy that characterized Mission Revival as the first manifestation of SCR held true for later iterations of the style. It was nearly all pure myth and symbol, mixed liberally in a cultural cauldron to produce architecture embodying the idea of a new and perfected Mediterranean.[12]

FIGURE 3 Union Pacific Depot, opened in 1904, Seventh and Pachappa, in the Mission Revival style, Riverside, ca. 1904

Lummis, Otis, and Miller's promotional success led the Santa Fe and Union Pacific Railroads to build depots at popular tourist destinations in variants of the Spanish Colonial Style. The Mission Revival, and its variant, Pueblo Revival, became the styles of choice in the late nineteenth and early twentieth centuries, to reinforce the idea that travelers were embarking on an exotic adventure to America's own Mediterranean, and the Southland's own Spanish past. When travelers descended at the Mission Revival Union Pacific Depot in Riverside, at once a modern train station and Moorish mirage of arches and four domed towers, their destination was often Frank A. Miller's Glenwood Hotel, California's Mission Inn (fig. 3).

As the Ramona Myth and related legends of the California Mission Period began to take root in the Inland Empire, the rampant popularity of the SCR they fostered in Riverside—founded in 1870 by radical abolitionist Judge John Wesley North, and a contingent of the Anglo northern middle class—made it the undisputed center of the "Mission Cult."[13] Engulfed in the 20,000 acres of Riverside navel orange groves, Miller's Glenwood Mission Inn comprised an unsurpassed, though contrived Spanish resort destination among those groves. In the heart of Riverside, the Mission Inn was the chief icon of the Mission Revival. Only a few blocks west of the Union Pacific Station, guests who wished to walk the short distance could promenade to the Inn under wisteria-bearing, faux-citrus branch pergolas of cast concrete; all the while growing more and more intoxicated on the fragrance of orange blossoms. Once there, they were greeted by a tile entry plaque featuring images of two Franciscan friars, flanking the proclamation, "Entre es su casa amigo, California's Mission Inn, Riverside." This confirmed that visitors had reached the ultimate in Spanish hospitality, since "Between us friend, this is your home, California's Mission Inn, Riverside."

Inspired by Miller's intense love affair with the Mission Cult, and designed by the leading architect of the Mission Revival, Arthur B. Benton of Los Angeles, the hotel opened with

FIGURE 4 Courtyard, New Glenwood Hotel, California's Mission Inn, ca. 1905. Arthur Benton, architect

a flourish in May 1903. President Theodore Roosevelt himself presided over the ceremony. Miller's Inn took Lummis' Mission Cult about as far as it could go, in Kevin Starr's view. Arthur Benton's design prowess, and his extensive knowledge of the architecture of California's Spanish missions made that possible. The Inn's "Mission" arches along Seventh Street, the faux-citrus branch Adirondack Stick Style pergolas that lined the U-shaped Court of the Birds, stucco walls, scalloped parapeted rooflines, Mission dormers, and Arts and Crafts "Mission" interiors, all exuded the Mission myth (fig. 4). Benton designed the early additions to the Mission Inn too, including the Cloister Wing, inspired by the Carmel Mission, helping Miller build the Mission Myth bigger each time the hotel expanded.

If visitors imagined themselves staying in an actual mission, the Master of the Inn seldom disabused them of it. He reinforced and enhanced the myth of the Inn by petitioning City Park officials for one of the late Eliza Tibbets' two original parent navel orange trees, and replanted it during Roosevelt's visit in 1903 in front of the Old Tea Room at the Inn, leaving the impression in the minds of tourists that it had been there all along. After all, citrus had been introduced to California at Mission San Gabriel, and perhaps the Inn had a hand in it somehow.

By the Inn's debut in 1903, Mission Revival structures were springing up throughout the downtown. Travelers from the Union Pacific Depot to the Inn passed two significant Mission Revival buildings within a block of the hotel. At the corner of Sixth and Lemon Streets, Arthur Benton's First Church of Christ, Scientist (Christian Science), ca. 1901, predated the Mission Inn by two years. Its exotic Moorish-Mediterranean contours inspired Miller to select Benton for expansion of his Anglo-style Glenwood Cottages into a Mission Revival extravaganza. On first sight, the Mission Revival church undoubtedly conjured up images in the minds of visitors of Byzantium and the Moorish architecture of North Africa and Southern Spain. The striking building was designed as an amalgam of Byzantine, Moorish, Classical, and Mission Revival.[14] Its north elevation features a giant classical order of Ionic columns and pedimented gable, capped by a red clay tile roof, supported by unfluted Ionic columns. Parishioners enter the building at the northwest corner through arched doorways, set at the base of a three-tiered Moorish tower. The sanctuary sits under a Byzantine dome, encircled at the base by clerestory windows. Though designed and built at the beginning of the new century, the church exemplified late nineteenth-century eclectic architecture (fig. 5).[15]

Immediately adjacent to the Inn on the east sat Franklin Pierce Burnham's Mission Revival Carnegie Library, which opened in July 1903. Burnham won a design competition for the library, and rendered it in a classic early Mission Revival, replete with arched portico entrance, marked by two Moorish towers, and curved parapeted Mission gable (fig. 6). Public outrage over the demolition of the library building in 1969 led, in large part, to Riverside's historic preservation program.

By 1913, Riverside also claimed standout residential Mission Revival structures. "Ridgecourt," the superb Mission Revival residence of Clinton Hickok, 3261 Strong Street, on a knoll situated along what became La Cadena Drive north of town, epitomized the form. Hickok, a wildly successful piano dealer from Toledo, Ohio, commissioned local architect G. Stanley Wilson to design the house and grounds, and

FIGURE 5 First Church of Christ, Scientist, Sixth and Orange Streets, eclectic early Mission Revival, built in 1901. Arthur Benton, architect

instructed the architect to design the patio in the shape of a grand piano, and the exterior in sophisticated Mission style. The Cresmer Manufacturing Company, one of Riverside's noted construction firms built the house with stucco finish over hollow clay tile, considered the latest in fireproof construction material. The front elevation features a central curved Mission parapet roof with flanking symmetrical arcaded bays. Wilson's room layout and details employed the fashionable Arts and Crafts interiors dominating California in 1913. Hickok turned the ten acres of his estate into a well-planted park-like setting in the California Style; with palms, cacti, yucca, California Pepper Trees, and a profusion of roses, and flowering shrubs. Most of the plants came from the United States Department of Agriculture, Gardens and Grounds Division, and were tended by an on-site gardener who lived in a back cottage (fig. 7).[16]

William Boyd, and his wealthy Bostonian bride Laura, bought the house from Hickok in 1917. Boyd later attained celebrity status as Hopalong Cassidy in movie and TV Westerns. Shortly after purchasing the property, William and Laura Boyd commissioned local architect Welmer P. Lamar to design $10,000 of improvements, including an expansive billiard room matching Wilson's original interiors, and a detached maids' quarters. In less than a year, the couple separated, with Laura returning to Boston and William to his rendezvous with Hollywood Westerns.[17]

Redlands' citizens watched in wonder in 1898 as the A. K. Smiley Public Library rose before their eyes. Dedicated on April 29 of that same year, and designed by local architect T. R. Griffith, the building resembled nothing else in town. Griffith's building evokes images of a far-away place, reflecting both the architect's vision of a California Mission and a Moorish castle, at once liberally entangled with the ornamentation and weightiness of the Spanish Romanesque (fig. 8).[18] The Library has a cruciform shape, marked at the entrance by a thick Italianate tower capped by a Moorish dome (fig. 9). Altogether, the building exhibits the eclectic combinations of both Islamic and

FIGURE 6 Mission Revival Carnegie Library, Riverside, built in 1903, demolished in 1970; ca. 1930. Franklin Pierce Burnham, architect

FIGURE 7 "Ridgecourt," the Clinton-Hickok Residence, ca. 1917. G. Stanley Wilson, architect

FIGURE 8 A. K. Smiley Public Library, Redlands, ca. 1898. T. R. Griffith, architect

Mediterranean motifs: it speaks the vocabulary of Moorish Mission in the front façade's arcade, flowing scalloped parapet over the arcade, its surface ornamentation, and battlement over the porte cochere. The recessed brick treatment in the contours of the arches, and the carved ornamentation on piers supporting the arches of the two porte cochere structures, however, declare Spanish Romanesque.[19]

Launched through the patronage of Albert K. Smiley, a wealthy easterner-turned-permanent resident of Redlands, and aficionado of the Mission Cult, the Library became an immediate icon of the rising Mission Cult in Redlands. Albert Smiley and his twin brother Alfred were leading members of the Mission Indian Federation, which included Frank Miller, Charles Fletcher Lummis, and other luminaries of the era, who dedicated their time to civic affairs built around the Spanish myth of Southern California. The Library's design was intentionally aimed at promoting the myth and its associated causes, including activism on behalf of furthering the station of Native Americans.

Redlands' most spectacular residential Mission Revival building is the mammoth Albert Burrage Mansion, "Monte Vista," 1205 W. Crescent Avenue, built in 1903. The mansion is on the scale of a major institution, more than a home. Charles Brigham of Boston drew the architectural plans and local architect Charles C. Coveney served as the on-site supervisor of construction. The *Redlands Daily Facts* reported:

> the style is Mission, of that of Christian, Spain, lacking the exterior ornamentation which characterizes the Moorish. The *Facts* reporter went on to describe the lavish "central portion of the first floor" as "occupied by a hall designed in the Pompeian style, with colonnades surrounding a fountain, with a skylight above, and the walls to be covered with

FIGURE 9 A. K Smiley Public Library Italianate entrance tower, with Mission-style scalloped parapet above posts

FIGURE 10 Hand-tinted postcard, Mission Revival Albert Burrage Mansion, Redlands, ca. 1903. Charles Brigham, architect

FIGURE 11 Santa Fe Depot, San Bernardino, eclectic Moorish Mission. W. A. Mohr, architect

rural decorations.... Two towers for observation purposes occupy the center of the main front. The stable is also Mission style with accommodations for half a dozen horses, a coach house and other rooms (fig. 10).[20]

Franz P. Hosp, Riverside landscape architect and nurseryman, designed and planted the grounds with Canary Island palms along West Crescent and more leading to the decorative one-hundred-step entrance. Floral arrangements followed designs Hosp had completed in Cañon Crest Park and Prospect Park.[21]

A few miles west of Redlands, the railway hub city, San Bernardino, produced one of California's largest and most iconic Mission Revival buildings. The picturesque behemoth is a dreamscape of "scalloped parapets, red tile roofing, deep overhanging eaves, domed bell towers, long arcades, quatrefoil windows, round arched windows, and balconies."[22] Designed by Los Angeles architect W. A. Mohr, the Atchison, Topeka and Santa Fe Railway Passenger and Freight Depot (Santa Fe Depot) at 1170 West 3rd Street, San Bernardino, opened in 1918. Mohr's design takes a strong horizontal orientation and asymmetrical composition consistent with the architectural vocabulary of the Mission Revival style, including added Moorish influence from four massive domes of the four towers on the center bay of the structure. Unlike many contemporary Mission Revival buildings, the Depot's massive central bay is derived directly from the classical order on the front façade of original Mission Santa Barbara, based on the work of the great Roman architect Vitruvius, consulted by the friars in the layout of the mission (fig. 11).[23]

Rise of the Mediterranean Revival, ca. 1914-1935

By the early 1920s, the original forms of the Mission Revival, which Gebhard called a nineteenth-century style, apparently no longer met the rising expectations and aspirations of many Southern California civic leaders and architectural critics. Disenchanted with indigenous styles based on the California missions, they began to clamor for more sophisticated European forms. In response, a new generation of California architects, educated in schools of architecture dominated by the Beaux-Arts style, and under professors who had trained abroad at École des Beaux-Arts in Paris, was ready to address these more Eurocentric requirements. The academic Beaux-Arts curriculum taught a rigorous method of design based on the philosophy and principals of the ancient Greco-Roman architects. Derived in large part from the work of Vitruvius, in his magnum opus, *De Architectura,* bible of the Roman Imperial builders, Beaux-Arts curriculum laid down the classical principles of architecture and engineering mathematics necessary to implement the universal principles undergirding Greco-Roman building theory. Certain regional architects, pursuing the Beaux-Arts model, made the grand tour to Europe, sometimes staying behind to study in Paris and to visit Spain and Italy where they made studies of ancient Greco-Roman and Renaissance buildings and landscapes. These architects made a smooth and quick transition to the Mediterranean Revival.[24]

In response to the rising demand for European derived interpretations of the California Style, the Mission Revival itself transformed into a more sophisticated, distilled and restrained form that better conformed to the new and dominant Mediterranean Revival.[25]

The precedent for this transformation had existed for some time in Southern California. Parallels between the region and the Mediterranean had been enunciated as early as 1891 by writer Charles Dudley Warner, who asserted that this American Mediterranean had "a more equable climate . . . than the North Mediterranean can offer."[26] Architects capitalized on the comparison to strip down their Mission Revival forms in favor of large buildings with unadorned surfaces, and less rococo ruffles and flourishes, often in contrast to the more exuberant iterations of the 1920s style.

Myron Hunt and Bertram Goodhue catalyzed the region's architectural transformation from Mission to Mediterranean Revival. Hunt's role came first in Riverside in 1914, with Goodhue second in San Diego in 1915. These two architects gave Southern California its signature California Style; the SCR in its mature form.

Hunt's Riverside project came first, and Frank Miller made it happen. In 1911, Miller's First Congregational Church building committee was charged with constructing a new church on Seventh Street, a half block east of Miller's Mission Inn. Considered key to implementing the new Riverside Civic Plan by Arthur Benton and Myron Hunt, Miller's committee sponsored a design competition for the church sanctioned by the American Institute of Architects. In a distinct break with the prevailing Mission style, competition rules specified a design in Spanish Renaissance. Three architects were invited to compete: Arthur Benton, Lester S. Moore, a noted Los Angeles architect of the Mission Revival, and the firm of Hunt and Grey of Pasadena. Designs of Benton and Moore mistakenly confused Mission Revival with the Spanish Renaissance, and

FIGURE 12 Spanish Renaissance Revival First Congregational Church, Riverside, opened 1914. Myron Hunt, architect

were disqualified. Hunt and Grey won the competition with an accurate Spanish Renaissance design, though they had parted ways prior to the award. It appears that Hunt's understanding of the difference reflected his European experience and Beaux-Arts training (fig. 12).[27]

Hunt stripped his Spanish Renaissance church of parapet walls, Mission dormers, and other Mission Revival and Moorish elements, applying classical principles of balance and harmony to establish the scale, mass, and rhythm of the structure. From there he drew on European Renaissance and Mexican Colonial models of cruciform churches for guidance on applied Churrigueresque detail. The mass of the Beaux-Arts church building features plain, unadorned walls "In a manner consistent with the precedent of eighteenth-century Spanish and Mexican examples of the Churrigueresque style," and "concentrated the cast stone surface ornament on . . . the three tiers and dome of the tower, the entrance arcade on the Mission Inn Avenue façade, and the first and second story window surrounds on the north wall of the northern transept."[28]

He raised the tall three-tiered Churrigueresque carillon tower at the east end of the nave, and decorated it with Spanish and Aztec iconography (fig. 13).[29] Local architect and engineer, Henry Jekel, engineered the tower and drafted to scale the artificial stone ornamentation for casting by the local contractor. His engineering skills are reflected in the church's reinforced concrete foundation and carillon tower.[30] Hunt's Spanish Renaissance Revival church opened in 1914 to great fanfare, including publication in *American Architect*, May 1914, and quickly won national acclaim. With it began the Mediterranean phase of the SCR in Southern California.[31]

The second and final push took place at Balboa Park, San Diego, in 1915, when renowned and flamboyant Arts and Crafts architect Bertram G. Goodhue—known in the East for his English Gothic Revival churches, and in the West as an expert in the historic Spanish Colonial architecture of Mexico[32]—completed the commission for the buildings of the Panama-California Exposition. Goodhue's California Building ignored Mission Revival elements completely in favor of exuberant Mexican Spanish Colonial church architecture, with liberal doses of Churrigueresque ornamentation. He raised a bombastic three-tiered tower as a beacon for fair goers, shrouded in elaborate Churrigueresque detail. Gebhard says that once Goodhue unveiled his exuberant Spanish Renaissance Revival California Building, the Mission Revival looked old-fashioned.[33]

Within a few years after the San Diego exposition, regional architects, builders, realtors, and materials manufacturers recognized the economic and cultural power of the rising style, and quickly acted to capitalize on it. The Spanish Colonial became synonymous with high standards of construction and advanced building materials—including structural concrete, stucco, tile, and wrought iron—as markers of a pervasive belief among the regional Anglo elite that European American civilization was superior to the native Mexican and Mexican American cultures of California. Often modern methods and materials cost more than traditional forms, and that differentiated between those who could and could not afford the construction methods or the style.[34]

In 1929 Riverside moved decisively to codify this mature phase of the SCR as the city's official architectural program. Planning commissioners hired City Planner Charles Cheney, designer of the Santa Barbara Civic Plan earlier in the decade, to redesign the existing Hispanicized Beaux-Arts Civic Center, by moving its focus fully to Mediterranean Revival. His resultant elaborate design proposed to remake the downtown Civic Center into a Spanish Plaza, replete with a new SCR library by Myron Hunt; and a replica of Seville's Giralda Tower, which Miller proposed to build at the southeast corner of the Mission Inn, at Seventh and Orange.[35] Although the Planning Commission, Frank A. Miller, and the Riverside Chamber of Commerce vigorously endorsed the Cheney Plan, it was scuttled by the 1929 stock market crash and the onset of the Great Depression.

Parts of the Cheney Plan, however, were already underway, including a new soldiers' memorial. When the Memorial and Municipal Auditorium went up adjacent to the Christian Science Church, the original Mission Revival style, a nineteenth-century style, was no longer dominant. The Auditorium, however, refused to stray far from it. Designed by Arthur Benton, and completed by G. Stanley Wilson, one of Riverside's best known architects of the

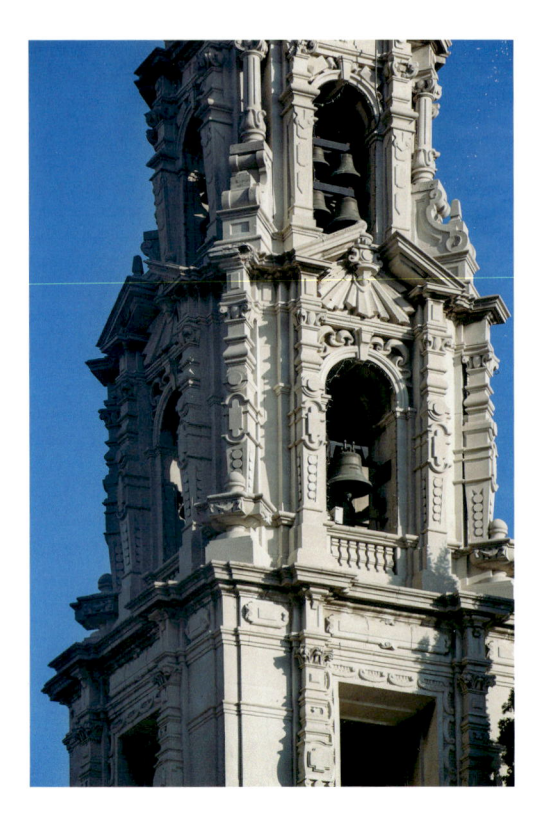

FIGURE 13 Detail of elaborate Churrigueresque ornamentation, carillon tower, First Congregational Church, Riverside

FIGURE 14 Hand-tinted postcard of distilled late Moorish Mission Revival Municipal Auditorium, Riverside, ca. 1935

twentieth-century, the Auditorium took a distilled Moorish Mission Revival form, while quietly featuring modern technical and engineering infrastructure. Built of unpainted steel-reinforced, slip-form concrete, the Auditorium abandons any pretense of replicating mission construction techniques, opting for a modern twentieth-century structure that exudes stability and power. Simultaneously, it intentionally retains medieval references from the Mission Santa Barbara that mark its Arts and Crafts bona fides.[36] The Auditorium's claim to Mission style resides exactly in Benton's application of modified medieval ecclesiastical form. The Auditorium reaches toward heaven in the mode of a restrained Gothic cathedral; adding Moorish flair by capping its front elevation high (west) and low (east) towers with brightly colored and glazed tiled domes, and window treatments and fixtures of wrought iron. The west elevation features a cloistered walk under red clay tile roof, á la Mission Santa Barbara, and a sunken patio, lending an ecclesiastical air to the form (fig. 14).

Adjacent to the Auditorium, Riverside's Mediterranean Revival YWCA, by noted California architect Julia Morgan, anchors the east end of the Civic Center. It features Spanish Renaissance, Palladian, and Beaux-Arts classical elements, all under a red clay tile roof. The interior is expressed in a matching vocabulary of eclectic Mediterranean Revival; supported by Ionic columns, with decorative Spanish Revival hand rails, and a formerly open air atrium.[37] The building has housed the Riverside Art Museum since the Art Association, successor to Frank Miller's 1910 Spanish Art Association, acquired it from the YWCA in the late 1960s, and was placed on the National Register of Historic Places in 1982 (fig. 15).

By 1935, Riverside's downtown Civic Center on Seventh Street (Mission Inn Avenue) featured a refined cluster of superb Beaux-Arts SCR public and religious structures, which updated and complemented its anchor—the now fully built-out Mission Inn. In 2009, the Seventh Street Historic District, hosting these buildings, was listed on the National Register of Historic Places as Seventh Street National Register Historic District, with most of the contributing structures previously individually listed. Those buildings include principally, among others, Myron Hunt's Churrigueresque Congregational Church, 1914; the adjacent Italian Renaissance/neoclassical (with Mission elements) Federal Post Office, 1914, by architects of the United States Treasury, James K. Taylor, Supervising Architect; the Churrigueresque Spanish Renaissance City Hall, 1924, by San Bernardino architect Howard Jones; the Riverside-Arlington Heights Fruit Exchange, ca. 1923, by G. Stanley Wilson; the Loring Theater and Office Building, remodeled from Richardsonian Romanesque to Mission Revival in 1918 by G. Stanley Wilson; and the Stalder Building façade, ca. 1926 by Wilson; and the distilled Mission Revival Fox West Coast Theatre 1929, 3801 Mission Inn Avenue (northwest corner of Mission Inn Avenue and Market Street) that anchored the west end of the Civic Center.[38]

Designed by the Los Angeles firm of Balch & Stanbery in a distilled twentieth-century Mission exterior, the Fox has a stucco finish over brick. The theater's corner entrance is marked by a three-story Mission tower at Mission Inn Avenue and Market Street. Arcades run the length of the east and south elevations, reminiscent of Mission Santa Barbara. Lavish interiors include massive stenciled beams in the mode of Southern Spain, lush carpets, and decorative iron hand rails that signal to movie goers they have arrived at a true California Style movie palace. During the 1930s and 1940s, the Hollywood film studios used the Fox Theater to preview future movies prior to

FIGURE 15   Riverside Art Museum, Mediterranean Revival YWCA building, Riverside, built in 1929. Julia Morgan, architect

FIGURE 16   Fox Theatre, Riverside, opened January 1929; ca. 1933. Balch & Stanbery, architects

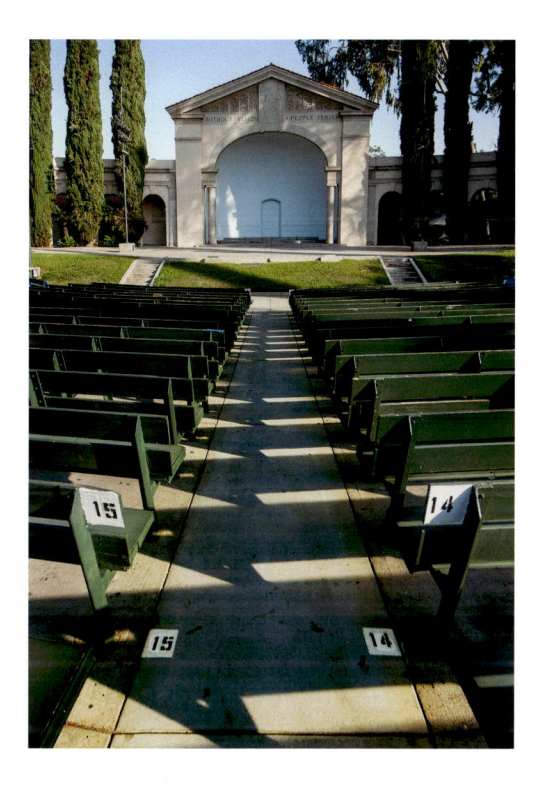

FIGURE 17   Redlands Bowl, Redlands, California. Mediterranean Revival amphitheater

FIGURE 18   Federal Post Office, Redlands, eclectic Spanish Colonial Revival with Moorish influences. G. Stanley Wilson, architect

editing and final release. Considered an ideal preview site, Riverside's Anglo Protestant movie-going demographic resembled small town America. Studios could gauge the reaction of this prime demographic within sixty miles of their base of operation. The theater previewed the first public showing of *Gone With the Wind*, among other "firsts." Recently, the city of Riverside completed a comprehensive restoration of the theater, converting it into the Riverside Performing Arts Center (fig. 16).[39]

Meanwhile, Redlands, San Bernardino, and Corona were remaking their own Civic Centers in the Mediterranean Revival. Redlands Music Association, in the mode of ancient Greece and Rome, brought an Amphitheater for the performing arts directly to the heart of the city in 1924. The Redlands Bowl took shape directly adjacent to the A. K. Smiley Public Library, Redlands' temple of learning, where the Bowl continues to serve as a venue for musical and theatrical performances offered gratis to the public. Florence and Clarence White funded and commissioned the stage structure as a gift to the City of Redlands in 1931, inscribing a phrase from Proverbs 29:18, "Without vision a people perish," across the frieze above the stage.[40] The Bowl is a contributing structure to the Smiley Park National Register Historic District (fig. 17).

Fresh off the International Wing of the Mission Inn, G. Stanley Wilson designed one of Redlands' most notable SCR civic structures in 1934, toward the end of the Revival's popularity. The Federal Post Office building, a stone's throw northwest of the Redlands Bowl, is "[c]onsistent with Wilson's idiosyncratic personal style," and "contains an eclectic mix of architectural styles. The predominant character of the building relates to (distilled) Mission style, but elements of the (Mediterranean) SCR are used extensively as well. The building does not conform to the symmetry typical of most post offices and, instead, relies on its irregular plan and massing to provide a picturesque" image of Moorish, Spanish and distilled Mission Revival.[41] Wilson made a Moorish-influenced octagonal polychrome tile-covered dome, topped with a concrete lantern, the prominent feature of the building (fig. 18).[42]

The California Theatre at West Fourth Street in downtown San Bernardino exudes the romance of SCR in its mature form. Fox West Coast Theatre Corporation opened the California with high fanfare in August 1928, bringing actors Janet Gaynor (1928's academy award winner) and Charles Farrell, who were starring in the opening night movie, *Street Angel*. The California was designed by prominent architect John Paxton Perrine in an ornate Spanish Eclectic style to match Fox's existing chain of opulent West Coast "movie palaces." Perrine was noted for designing theaters in rococo SCR. The Theatre's Fourth Street front-facing façade is an exuberant mix of elements.[43] The overall look is exotic and highly romantic, leaving visitors with the image of a West Coast movie palace during the "Golden Age of Hollywood." The California Theatre is listed on

FIGURE 19 California Theatre, San Bernardino, in exuberant Spanish Colonial Revival. John Paxton Perrine, architect

FIGURE 20 Corona High School, Corona, Mediterranean Revival, built in two phases, ca. 1923, 1931. G. Stanley Wilson, architect

FIGURE 21 Palm School, Riverside, Mediterranean Revival. G. Stanley Wilson, architect

FIGURE 22 Corona Theatre (Fox), Corona, ca. 1929. Carl Boller, architect

the National Register of Historic Places, and for several years has served as San Bernardino's performing arts center. The California retains its Wurlitzer style-216 theatre organ (Opus 1850) in its original location, adding to the period ambience (fig. 19).[44]

The citrus belt town of Corona, located south of Riverside, joined the 1920s move to Mediterranean Revival with two choice buildings: the 1929 Fox Corona Theatre on Sixth Street, and Corona High School, built in two phases, 1923 and 1931, West 6th Street. Designed by the firm of G. Stanley Wilson, today the school serves proudly as the Corona Civic Center in the Corona Historic District. The original campus occupied an entire block on West Sixth Street, extending all the way to Third Street. Riverside's Cresmer Manufacturing Company built the 1923 reinforced concrete Administration Building and Auditorium and the Domestic Science Building (Home Economics). In contrast to the later Atrio of St. Francis at the Inn, Wilson rendered the school in a restrained combination of Beaux-Arts Spanish and Italian Renaissance style (figs. 20–21).[45]

Corona landed a Fox West Coast Theatre in August 1929. Carl Boller, of Boller Brothers, Architects, ranked among the top five theater designers in the United States, designed the Theatre. Corona Theatre exhibits a complex plan "reminiscent of the varied roof forms of Spanish villages."[46] The L-shaped SCR structure rises two stories along the front elevation, and three stories at the rear. A two-story square Italianate tower marks the southwest corner. The Corona movie palace seated less than the theaters in the larger cities of Riverside and San Bernardino, but movie-goers felt it matched them in style and comfort (fig. 22).

Residential Design: Courtyard Houses in the Spirit of Andalusia

The Courtyard House and
the Castle in Riverside
Gebhard said that, "[i]n the 1920s, the whole of Southern California seemed about to be transformed into a new, much improved, Mediterranean world." Residential architecture appeared to validate the claim.[47] Early residential design in the Revival drew its inspiration from the ubiquitous Greco-Roman courtyard houses present across the Mediterranean regions of Europe and North Africa. The courtyard house, arranged inward around an open-air center court, usually consisted of a one or two-story residence constructed of masonry, with red clay barrel-tile roof. These homes appear to spring from the garden, with a strong emphasis on indoor-outdoor living. Especially in California, Mediterranean Revival versions of the courtyard house facilitated an easy passage to the

courtyard and the immediate outdoors. Often, they were arranged to offer a vista of mountains and surrounding terrain intended to evoke the ambience of Spain or Italy.⁴⁸

The Roman *villa suburbana*, designed as a retreat from the city, usually made a terraced landscape the dominant feature of the design.⁴⁹ By the end of the 1920s, Riverside hosted superb examples of the Mediterranean Revival iteration of the *villa suburbana*, in particular, the Fred Krinard and Harry Hammond Estates. Both were located on Victoria Hill in the exclusive Country Club Park, overlooking the Victoria Club golf course. The houses and grounds were designed by Riverside's Arts and Crafts architect and Buffalo, New York, transplant Henry L. A. Jekel.⁵⁰ By the mid-1930s, both these estates had earned National Yard and Garden Awards for their California Style terraced landscapes. The Hammond House, in the mode of a Mediterranean villa, epitomized the *villa suburbana*. The Krinard Estate too exemplified the form, especially in its terraced and thoughtfully planted front yard (fig. 23). It has an eclectic façade featuring a combination of Andalusian Farmhouse Vernacular, Monterey, and sophisticated Moorish detail, with Deco art glass and wrought-iron light fixtures. This amalgam reveals that Jekel shared the common penchant of his peers to assemble specific features from an array of Italian, Spanish, Moorish, and Southwestern influences, adapting them to American house types reinterpreted in the California Style for Southern California.⁵¹

Most Mediterranean Revival houses built in the early stages of the design phase, 1919–23, were inspired by romanticized visions of Spanish castles, particularly the famous Moorish citadels of Alhambra and Alcazar, both with inner courtyards. Charles W. Benedict, a wealthy financier from New York, commissioned Henry L. A. Jekel to build an ode to Alhambra on the southeastern heights overlooking Riverside, a location that reminded him of Alhambra. Benedict named the Spanish castle "Castillo Isabella," after his mother.

FIGURE 23   Entrance tower, Krinard Residence, Riverside, Mediterranean phase Spanish Colonial Revival Monterey, ca. 1928. Henry L. A. Jekel, architect

FIGURE 23A   Living room, Krinard Residence, featuring massive Spanish fireplace, art glass windows, and hand-hewn stenciled beams, ca. 1933

The millionaire's abode was a Moorish Revival tour de force, featuring a sixty-foot tower, a massive hand-hewn front door crafted in Spain, and a miniature version of the Court of the Lions at Alhambra, without the Lion Fountain. The interior matched the exterior in its monumental scale, scope, and detail. Although its size and castle-like arrangement of rooms made it awkward as a house, Castillo Isabella was well suited as a stage set for Hollywood parties, of which there were many during Benedict's time (fig. 24).[52]

Andalusian Farmhouse Vernacular

By the mid-1920s, however, the preference for the Iberian castle among Southern California architects and their clients had been supplanted by the reinterpreted vernacular farmhouse architecture of Andalusia (Southern Spain). The sudden shift in taste began on Middle Road in Montecito, California, in 1918, when newly arrived artist and architect George Washington Smith designed his personal residence and studio on a plan directly inspired by the residential vernacular architecture of Southern Spain. When he married Mary Greenough in 1912, the couple moved to Europe for three years, where they traveled through France, Italy, and Spain visiting museums, architectural monuments, and art galleries. Both were deeply impressed by the farmhouses and landscapes of Andalusia. The outbreak of World War I in 1914 brought them back to the United States, and to Santa Barbara, where Smith intended to establish himself as a California plein air artist.

FIGURE 24 Benedict Castle, "Castillo Isabella," Riverside, eclectic Moorish Mission Spanish Colonial Revival, Iberian castle phase, ca. 1921–31. Henry L. A. Jekel, architect

Smith's home and studio featured:

> the plain, stuccoed façade of a Spanish house, with few window openings and a simply molded, deeply set doorway emphasized by a balcony set above it.... The arrangements of the elements of the façade, placement of the doorway and windows, and the contours of the chimney and broken rooflines present an asymmetrical composition. This lack of symmetry, typical of farmhouses in Southern Spain, informs the irregular massing of volumes that is characteristic of Smith's designs.[53]

Patricia Gebhard reports that Smith's house on Middle Road was an instant success. The architect himself wrote in 1926, "this little house was practically the start of the Spanish Revival in Southern California...."[54] Smith's plans were published and republished in California and all across the country, beginning in 1920 with an article in *Architectural Forum*, arguing that Smith's design was a "germ of hope for future California architecture."[55]

Regional architects quickly picked up Smith's ideas, and his modified Andalusian Farmhouse Vernacular became the inspiration for much of the Mediterranean Revival residential architecture of the region. These houses were designed as sculptural volumes, tightly, but informally knit to the house site, and focused on indoor-outdoor living via patios, courtyards, verandas, and pergolas. Thick walls, finished in stucco and punctuated with deeply recessed door and window openings, lent an air of authenticity to buildings that were normally wood-frame structures, made to imitate masonry. Common features included red-clay barrel-tile roofs, Juliet balconies with iron railings and ornamentation, front entrances frequently recessed into deep wall openings, heavy wood doors with view port windows girded by wrought-iron cages, and decorative medieval-style metal strap hinges and plaster or tile door surrounds.

In Riverside, architect Henry L. A. Jekel excelled in rendering SCR houses and larger structures in Smith's picturesque variant. Jekel understood Beaux-Arts Classical style, but was an Arts and Crafts architect through and through. Of the same age and architectural temperament as G. W. Smith, Jekel constantly cast an architectural eye toward the primitive or rustic, employing medieval elements in his design while using the latest engineering technology and materials. "Lion Head," on Rumsey Drive, the 1926 eclectic SCR work of art for Laguna Beach real estate mogul, Howard G. Heisler, incorporated neo-medieval Spanish elements that matched the Krinard Estate on Victoria Hill. Both estates reflect Jekel's consistent resort to the medieval, including rusticated wrought-iron fixtures and details, balconies, and art glass windows, evoking the venerability of medieval stained glass. His romantic revival designs invoked medieval castles, with their great halls (living rooms), open-beamed cathedral ceilings, and loggias.[56] Jekel also incorporated many of these elements in his plans for middle-class houses in the Wood Streets, Riverside's well-known streetcar suburb, situated around New Magnolia Avenue.[57]

Jekel, who adhered closely to Arts and Crafts principles, generally retained the "natural" gray tone of stucco exteriors for his house designs, unless the client insisted on a different treatment. The architect always tried to stay true to his materials, leading him to oppose painting structural concrete exteriors, and lauded G. Stanley Wilson for not painting the

FIGURE 25 Elijah Parker Residence, Riverside, Spanish Colonial Revival Monterey, mature Mediterranean phase, with Andalusian Farmhouse vernacular influences, ca. 1927. Robert Spurgeon, Jr., architect

FIGURE 26 Promotional advertisement, Casa de Anza Hotel, Riverside, Spanish Colonial Revival, mature Mediterranean phase, ca. 1926. G. Stanley Wilson, architect

reinforced slip-form concrete of the International Wing of the Mission Inn.[58]

Robert Spurgeon's houses, on the other hand, cleave to classical principles of balance and harmony, traits taught by the Beaux-Arts–driven American university schools of architecture, following the Parisian [or European] model. His 1926 plan for five superb houses at Riverside's western entrance never materialized, but five others continue to anchor the southern slopes of Little Mount Rubidoux, in the Mount Rubidoux Historic District. The terrain and distance from the developed urban core inspired Spurgeon to design there an enclave of *villa suburbana*. The first, a Mediterranean Revival villa, was designed in 1923 for his parents, using a combination of Italian and Spanish motifs to create an architecture that was quite romantic, but nonetheless resonates with his rigorous Beaux-Arts training.[59]

Spurgeon's hallmark 1927 Elijah Parker house on Ladera Lane reflects Andalusian and Monterey Colonial influences, superbly executed by Spurgeon (fig. 25). His SCR house at 5110 Magnolia Avenue clearly features his Beaux-Arts interpretation of the idiom, with its classic balance of projecting and receding elements on the front elevation. His most expensive houses in Riverside, including "Casa de Arroyo," overlooking Victoria Club golf course, and the $40,000 Roberts Leinau residence on Pachappa Hill, demonstrate Spurgeon's understanding of the Spanish Renaissance, and its purposeful and restrained application of ornament to otherwise unadorned flat surfaces. These houses presaged "Oaklodge," the Montecito Mediterranean courtyard mansion designed for Spurgeon's sister and her wealthy husband in 1930.[60]

G. Stanley Wilson's noteworthy Mediterranean Revival houses include "Casa de Anza Hotel," Landmark no. 85 of the City of Riverside (fig. 26). Its long evolution began as Wilson's own residence before being moved to Market Street, where he expanded it and turned it into one of Riverside's finest uses of SCR in commercial residential design. The Casa de Anza is a refined, mature, and well-articulated variant of the style, which incorporates red-clay barrel-tile roof, Spanish-style oriel bays, arcades, and elaborate glazed Malibu and Catalina tile door surrounds and interior tile details.[61] It demonstrates Wilson's mastery of the Mediterranean Revival in multiple occupancy residential architecture, stemming from the firm's intimate involvement with the design of the progressive phases of the National Historic Landmark Mission Inn, 1913–32.

San Bernardino's West 25th Street Historic District, once known as Doctors Row because so many local physicians lived on the street, represents an entire subdivision constructed in the reinterpreted Andalusian SCR, incorporating middle-class and high-end residences. The district consists of thirty-six homes on two city blocks (fig. 27). Fourteen of the one and two-story houses were built by contractor E. A. Anderson in 1928, scaled to suit mid-to-high 1920s budgets. Taken together, these houses speak the vocabulary of the 1920s California Style. Considered a premium home site choice of the area's up and coming, Mediterranean mansions also rose around the Arrowhead Country Club in north San Bernardino, although these houses strayed from Andalusian Farmhouse Vernacular in favor of more opulent Mediterranean variants. One Mediterranean Revival mansion on Parkside Drive is a prime example. The 1930 house has long rectangular east (front) elevation, which includes three distinct bays, containing Spanish Renaissance, Italian Renaissance, and Monterey Colonial features.[62]

Redlands features high-end modified Andalusian SCR housing stock in the Garden Hill Historic District, especially the outstanding Cecil Brashears house, better known as the Leslie Harris house, on Garden Hill. Built by a Mexican master of adobe construction, who apprenticed in Mexico, the Harris house is a rare example of a true adobe SCR structure built in the mature phase of the style, rather than a faux version built with modern materials, balloon-frame construction, and methods to imitate the real thing. Though there are many other mid-to-large Mediterranean houses in Redlands, the side-by-side homes of the McCulloch brothers, Robert and Hugh, on Canyon Road (ca. 1924 and 1926), illustrate a commitment to the California Style among the city's Anglo elite. The McCulloch brothers

FIGURE 27 West 25th Street Historic District, San Bernardino, looking north with Spanish Colonial Revival houses, ca. 1926

FIGURE 28 Early Spanish Colonial Revival Veterans Bungalows, Redlands, ca. 1921

FIGURE 29 Robert McCulloch Residence, Redlands, Spanish Colonial Revival, Mediterranean phase, ca. 1926. W. E. Rabbeth, architect

made their fortune in Cuba on the family coffee, and later sugar plantation, where they observed Spanish Colonial architecture for several years, adopting an American version of it once in Redlands (fig. 28).

"Bermejal" (Red Clay), the Robert and Kathleen McCulloch home, sat at the terminus of the Redlands Electric Railway Line, directly across from the Redlands Country Club (fig. 29). Architect W. E. Rabbeth designed the two-story Mediterranean Revival residence and Garret Huizing built it for $16,000.[63] Hugh McCulloch and his wife used the same architect and builder when they constructed their adjacent home in 1926. The *Redlands Daily Facts* called it "One of the most attractive homes ever erected here." It includes "a dull red Spanish clay roof, cross gabled, in three shades and exterior beam work in natural redwood. The exterior is finished in three coats of cement stucco. "Walled courtyards surround the rear of the house and are entered through plain arched doorways and multi paneled doors,"[64] in the style of Mediterranean courtyard houses.

A wealthy Corona citrus grower, Ryland A. Newton built an impressive two-story, red-tile-roofed, white stucco grove house south of downtown Corona in 1931. The house is replete with the signature elements typical of the style—red clay barrel-tile roof, porte cochere with pedimented gable, second floor French doors and balcony, along with a massive eight-foot wooden Spanish-style recessed front door. Newton's house made a strong statement for the popularity of the SCR among the inland area's prosperous citrus grower elite.[65]

Marion Featherstone built a middle-class version of the style in 1940 on South Main Street. Constructed of reinforced structural concrete, on a raised concrete foundation, the Featherstone house occupies a corner lot and features a symmetrical, half-turret corner entrance, accessed by a heavy wooden door with quoined surround. A large bay window with multi-paned, leaded art glass stands out on the east façade. The interior reflects the romance of the SCR, including high-coved ceilings, a curved staircase with hand-forged wrought-iron railings, hardwood floors, colorful glazed bathroom tiles, hand-painted wood

FIGURE 30 Casa Palmeras Apartments, Palm Springs, in refined Spanish Colonial Revival, Andalusian Farmhouse vernacular influences, ca. 1928. Paul R. Williams, architect

paneling, and romantic Spanish courtyard. These two disparate houses demonstrate how the California Style permeated all levels of Inland Empire residential architecture during the 1920s and 1930s.[66]

The Andalusian inspired, reinterpreted SCR began slowly in Palm Springs and surrounding desert communities, but surged in the mid-1920s. The arrival of Hollywood celebrities and wealthy Easterners, bent on making the desert their winter home, drove custom housing and commercial starts higher and higher. The desert communities are dotted with works by Southern California's most noted architects of the California Style, including from the city itself. Palm Springs Preservation Foundation works hard at preserving and promoting Desert Spanish Colonial, and publishing books about Desert Spanish Revival architecture (fig. 30).[67]

Conclusion

The Great Depression brought austerity to Southern California architecture, and with it the waning of the SCR, with its penchant for opulence and elaborate detail. Moreover, the rising tide of modernism—represented in Southern California by R. M. Schindler, Richard Neutra, Frank Lloyd Wright, and others—tended to make romantic buildings appear old-fashioned. Modern Movement architecture seemed right for a postwar nation looking to the future for its inspiration, not to the past.

Though it receded behind Modernism, the SCR nonetheless continued to hold its grip on Southern California. It returned after the war in the altered "skin" of Modern Movement buildings. G. Stanley Wilson's Riverside First Christian Church at Jurupa and Brockton materialized as a modern Spanish Revival form, red tile roof and all (fig. 31). The second iteration of St. Anthony's Roman Catholic

FIGURE 31 First Christian Church, Riverside. Spanish Colonial Revival in modern form, ca. 1952. G. Stanley Wilson, architect

Church in Riverside's Casa Blanca neighborhood reappeared as a modern abstracted SCR structure. St. Catherine's Roman Catholic Church, First Methodist, and other churches in Riverside readily joined the trend. By the 1970s, as Patricia Morton points out, modern reinterpretations of the SCR, high and low, had returned to Southern California with a vengeance.

Finally, by the late 1990s, the 1920s SCR itself had regained favor with home-buyers. In Riverside, by 2015, that fact resonated in the million-dollar price tag on Jekel and Spurgeon houses in the Mount Rubidoux Historic District and Country Club Park. On the heels of the trend, a few Southern California architects designing for wealthy clients returned to original 1920s SCR for inspiration, much as the architects of that era had looked to the Middle Ages and the European Renaissance for theirs. By the second decade of the twenty-first century, the reinterpreted SCR, in its high and low forms, had again become the architecture of inland Southern California.[68]

Notes

1. Dydia DeLyser, *Ramona Memories: Tourism and the Shaping of Southern California*, (Minneapolis: University of Minnesota Press, 2005), ix–xxiii. Kevin Starr, *Inventing the Dream: California Through the Progressive Era*, (New York: Oxford University Press, 1985), 65–91.
2. Patricia Gebhard, *George Washington Smith: Architect of the Spanish Colonial Revival*, (Salt Lake City: Gibbs-Smith, 2005), 30–34. See Jay Belloli, Jan Furey Muntz et. al., *Johnson, Kaufman, Coate: Partners in the California Style*, (Santa Barbara: Capri Press, 1992), op. cit., for a thorough discussion of three noted architects of the era who viewed the SCR as the "California Style."
3. Albert S. Fu, "Materializing Spanish Colonial Architecture: History and Cultural Production in Southern California," *Home Cultures*, 9 (February 2012): 150. The SCR remains a powerful hegemonic force in Southern California. Today's versions of the style persist in shaping the ubiquitous vernacular architecture of the region.
4. David Gebhard, "The Spanish Colonial Revival in Southern California, 1895–1930," *Journal of the Society of Architectural Historians* 26, no. 2 (May 1967): 131–47; and Gebhard, "The Myth and Power of Place: Hispanic Revivalism in the American Southwest," in *Architectural Regionalism: Collected Writings on Place, Identity, Modernity, and Tradition*, ed. Vincent B. Canizaro (New York: Princeton Architectural Press, 2007), 194–203.
5. Starr, *Inventing the Dream*, 65–91. See the interview with Bertram Goodhue, "One Great Idea in Building Emphasized by Noted Architect," *San Diego Evening Tribune*, n. d., 1916, RMM Collection. Intense antagonism toward the Catholic Church, and Catholics themselves notwithstanding, Anglo middle and upper classes of the nineteenth and twentieth centuries were also influenced by the Gothic Revival and the Arts and Crafts Movement. Both these movements looked to the medieval period for their models of piety and social reform, rather than the classical period of Ancient Greece and Rome. They argued that the inner motivation and hand-craftsmanship of the medieval period was truer to the human spirit than the search for rationality and balance of the Classical Era. Hence, these reformers could adapt Catholic church forms and icons to their purposes without adopting Catholic theology and practice. This convenient rationalization enabled Anglo appropriators of the Spanish Myth to do the same in the 1920s.
6. Starr, *Inventing the Dream*, 70–100; and DeLyser, *Ramona Memories*, ix–xxiii.
7. H. Vincent Moses, "'The Orange-Grower is not a Farmer:' G. Harold Powell, Riverside Orchardists, and the Coming of Industrial Agriculture, 1893–1930," *California History*, (Spring 1995): 22–37. Frank Miller pronounced himself "Master of the Mission Inn," and it stuck. Per the 1920 U. S. Census, Riverside had a population of 19,341; increasing to 29,695 in 1930, driving a substantial growth in Anglo middle and upper class housing stock.

8   Gebhard, "The Spanish Colonial Revival," 131–47.
9   Quoted in Starr, *Inventing the Dream*, 85.
10 Jocelyn Gibbs, Nicholas Olsberg, et. al., *Carefree California: Cliff May and the Romance of the Ranch House* (Santa Barbara: Art, Design & Architecture Museum, UCSB, 2012), 24–26.
11 See Starr, *Inventing the Dream*, 70–100; DeLyser, *Ramona Memories*, 161–87; and Vincent Brooks, *Land of Smoke and Mirrors: A Cultural History of Los Angeles,* (New Brunswick: Rutgers University Press, 2013), 29–54.
12 Gebhard, "The Spanish Colonial Revival," 132.
13 Starr, *Inventing the Dream,* 65–91.
14 See http://www.christianscienceinriverside.org/History.php for a brief history of the First Church of Christ, Scientist, Riverside, 3606 Lemon Street. See also City of Riverside Historic Property Inventory for the Church at http://olmsted.riversideca.gov/historic/ppty_mtp.aspx?pky=3925.
15 Ibid.
16 Esther H. Klotz and Joan H. Hall, *Adobes, Bungalows, and Mansions of Riverside, California Revisited*, (Riverside: High Grove Press, 2005), 236. *Riverside Daily Press*, September 11, 1913. Significant additional Mission Revival houses in the city included citrus magnate Martin Chase's mansion at 5145 Myrtle Avenue off Victoria Avenue, overlooking Chase's sixteen-hundred acres of prime navel orange groves and the Victoria Club and golf course; the Hayes-Pattee house, ca. 1903, at 3611 Mount Rubidoux Drive downtown; and the home of noted local entrepreneur and politician D. C. Twogood, at 3410 Prospect, ca. 1900, also in the downtown area.
17 *Riverside Daily Press*, April 28, 1917, and May 28, 1917.
18 For online sources about the Smiley, see http://focus.nps.gov/pdfhost/docs/NRHP/Text/76000513.pdf; and https://en.wikipedia.org/wiki/A._K._Smiley_Public_Library; and http://ohp.parks.ca.gov/ListedResources/Detail/994.
19 Phyllis Irshay, National Register of Historic Places Inventory-Nomination Form, *A. K. Smiley Public Library*, December 12, 1976.
20 Redlands Area Historical Society, 2010 Heritage Award Recipient, 1205 West Crescent Avenue, Redlands. See also http://www.burragemansion.org/.
21 http://www.burragemansion.org/; Tom Atchley, *The Burrage Mansion*, (Redlands: The Redlands Area Historical society, 2012), offers the best definitive history of the Albert Burrage Estate.
22 http://www.waymarking.com/waymarks/WMQW4T_Atchison_Topeka_and_Santa_Fe_Railway_Passenger_and_Freight_Depot_San_Bernardino_CA .
23 Vitruvius dedicated his magnum opus, *De Architectura, (On Architecture),* to his patron Caesar Augustus, and it became the guidebook(s) for the Roman Imperial building program during the Pax Romana. It is considered the most important original source on Roman architecture, engineering, and city planning extant from the period. The Renaissance brought it back into favor in Europe, and subsequently in the Spanish Colonies of the New World. For a look at the station, see https://en.wikipedia.org/wiki/Santa_Fe_Depot_%28San_Bernardino%29; and for a short history by Amtrak, see http://www.greatamericanstations.com/Stations/SNB.
24 See Gebhard, "The Spanish Colonial Revival," 136, who surmised, "the simple life was giving way to the affluent life of the 1920s." See Mark Gelenter, *A History of American Architecture: Buildings in Their Cultural and Technological Context* (Hanover: University Press of New England, 1999), 175–76.
25 Gebhard, "The Spanish Colonial Revival," 136. Romy Wyllie, "A Simpler Architecture," *Bertram Goodhue: His Life and Residential Architecture*, 130–41.

26 Charles Dudley Warner, quoted in Lauren Weiss-Bricker, *The Mediterranean House in America* (Abrams: New York, 1998), 6.

27 The extensive announcement of Hunt and Grey's winning design appeared in *Southwest Builder and Contractor* 23, October 15, 1910, including the initial illustration of the proposed church, showing its north elevation. See also "Congregationalists Accept Church Plans," *Riverside Morning Enterprise*, November 5, 1911; and "Beautiful Church is Accepted by Architect," *Riverside Enterprise*, Tuesday, December 6, 1913.

28 Ibid. Named for the seventeenth-century Spanish architect Jose Benito de Churriguera, Churrigueresque ornamentation reached its ultimate expression under Mexican Colonial architects of the eighteenth and nineteenth-centuries.

29 Janet Tiernan and Lauren Weiss Bricker, *First Congregational Church of Riverside*, National Register of Historic Places Inventory-Nomination Form 10-900 (Rev. 10–90), March 6, 1997: "For over eighty years (one hundred and two years in 2016) the highly visible . . . tower has served as an urban 'anchor,' signaling the entrance into Riverside's downtown via the Seventh Street/Mission Inn Avenue corridor. The church building and its generously landscaped linear forecourt perpetuated the image of Riverside as a Mediterranean city."

30 See also Jay Belloli et. al., *Myron Hunt (1868–1952): The Search for a Regional Architecture* (Los Angeles: Hennessey and Ingalls, 1984) for the argument that Hunt's 1914 First Congregational Church of Riverside launched for Phase Two SCR, by introducing the Spanish Renaissance Style to Southern California; and introduced it to the nation in the *American Architect* (May 1914).

31 Mr. Myron Hunt, Architect, "First Congregational Church, Riverside California," *The American Architect* CV, no. 2005 (May 27, 1914).

32 See Wyllie, *Bertram Goodhue*, op. cit. Goodhue thought of himself as an Arts and Crafts architect and an adherent of the philosophy of John Ruskin, William Morris, and the American Arts and Crafts Movement, choosing to select designs based on the site and region where his projects were located. In this way, he argued, his buildings rose out of the demands of the site and the locale, not out of universally applied principles to be imposed on the site by the architect.

33 Gebhard, "The Spanish Colonial Revival," 136–37.

34 Fu, "Materializing Spanish Colonial Architecture" 149–71. Fu looks at the builders and sellers of Southern California architecture of the 1920s, and the manufacturers of SCR building materials "to assess the material, economic, and cultural processes behind the making of Southern California's 'dream' house," 150. See also Harris Allen, "Spanish Atmosphere," *Pacific Coast Architect* 29 (May 1926): 5–7; and Tomas Almaguer, *Racial Fault Lines: The Historical Origins of White Supremacy in California* (Berkeley: University of California Press, 1994), who analyzes the social construction of race in California and the Southwest, and demonstrates how the dominant Anglo elite sought to differentiate itself from the working classes, and people of color; especially Mexicans and Mexican Americans. Ironically, the dominant elite made the SCR its own, as part of this effort, and a conscious marker of Anglos belief that European American civilization was superior to the native Mexican and Mexican American cultures of California.

35 Charles H. Cheney, *Recreation, Civic Center, and Regional Plan: Riverside, California* (Riverside, CA: Riverside City Planning Commission, 1929), op cit.

36 Mark H. Rawitsch, National Register Nomination Form, *Riverside Municipal Auditorium and Soldier's Memorial Building*, October 18, 1977. The Arts and Crafts Movement emerged in late Victorian England in response to rampant industrialization and the machine-driven diminution of labor. Based in part on the works of Pugin, Ruskin, and Morris, the movement sought to recapture the handcraftsmanship and pre-capitalist

forms of culture and society that the practitioners of the movement believed had existed in the Middle Ages, especially among the cathedral builders. See http://www.metmuseum.org/toah/hd/acam/hd_acam.html. Benton and other Arts and Crafts architects, in both Europe and America, were influenced by a romantic view of the medieval period, rather than the classical architectural principles of Greco-Roman philosophers and architects. They followed their medieval muse well into the new century.

37 City of Riverside Historic Property Inventory, Riverside Art Museum, http://olmsted.riversideca.gov/historic/ppty_mtp.aspx?pky=6963 for City Landmark record; and see National Register Nomination, Old YWCA, 1982. Other downtown civic center structures included the Riverside-Arlington Heights Fruit Exchange, 1923, by G. Stanley Wilson; the Stalder Building, ca. 1929, by G. Stanley Wilson; and the Loring Building, remodeled to Mission Revival in 1918. Away from the downtown, the University Heights Junior High School and Palm Elementary School, designed by architect G. Stanley Wilson, and Community Hospital, by Myron Hunt, expressed the popularity of variants of the SCR in civic structures.

38 City of Riverside Historic Property Inventory, Seventh Street NR Historic District, http://olmsted.riversideca.gov/historic/dist_mtp.aspx?dky=22.

39 City of Riverside Historic Property Inventory, Fox Performing Arts Center, http://olmsted.riversideca.gov/historic/ppty_mtp.aspx?pky=6978.

40 See https://en.wikipedia.org/wiki/Redlands_Bowl; and http://en.wikipedia.org/wiki/Redlands_Bowl#/media/File:Redlands_Bowl,_Redlands_CA.jpg.

41 David Gebhard and Robert Winter, *Redlands Main Post Office*, National Register of Historic Places Inventory Form for Federal Properties, December 27, 1984, at http://focus.nps.gov/pdfhost/docs/NRHP/Text/85000135.pdf.

42 2012 Heritage Award Recipient, The Redlands Area Historical Society, Inc., *U.S. Post Office Building*, 201 Brookside Avenue, 1934. Mediterranean Revival civic buildings of the era within Smiley Park include the Italian Renaissance City Hall and the SCR First Baptist Church, among others.

43 James L. Mulvihill, PhD, National Register of Historic Places Registration Form, *California Theatre of Performing Arts*, July 20, 2009.

44 Ibid.

45 Richard and Mary Winn, National Register of Historic Places Register Form, *Corona High School*, December 2004, in the Corona Public Library Heritage Room.

46 National Register of Historic Places Inventory-Nomination Form, *Corona Theatre*, 1991, in the Corona Public Library Heritage Room.

47 Gebhard, "Wood Studs, Stucco, and Concrete: Native and Imported Images," in *On the Edge of America: California Modernist Art, 1900–1950*, ed. Paul J. Karlstrom (Berkeley: University of California Press, 1996), 141.

48 Lauren Weiss Bricker, *The Mediterranean House in America*, (New York: Abrams, 2008), 8–9.

49 Ibid, 9.

50 For Hammond Estate, see the *Riverside Daily Press*, Wednesday Evening, June 20, 1928, 2; and *Riverside Daily Press*, February 11, 1936, 7, regarding the National Yard and Garden Prize. For the Fred Krinard Estate, see *Riverside Daily Press*, Wednesday Evening, May 23, 1928: "The beautiful new home of Mr. and Mrs. Frederick W. Krinard on the crest of Victoria hill is of typically Spanish architecture and is built to fit the contour of the eminence. The site is attractively terraced and landscaped and presents a pretty picture from Victoria Avenue. The home is set amidst tall, graceful palms and has a background of lovely eucalypti." Krinard Estate won a First-Place National Yard and Garden Prize in 1934 for the terraced landscape, and was featured in National Geographic, and in landscape nursery ads.

51 WeissBricker, *The Mediterranean House*, 7.
52 For a brief history of "Castillo Isabella," Benedict Castle, see Klotz and Hall, *Adobes Revisited*, 259–63, and Ursula Vils, "Everything but the Moat: Castle Benedict Has Address, History," *Los Angeles Times*, November 28, 1969. Thelma Perrin Davidson's scrapbook about the castle is comprehensive, and has been digitized by the authors. Benedict's widow said that the Warner brothers, various starlets, and Bob Hope were particularly fond of the castle.
53 Patricia Gebhard, *George Washington Smith*, 11.
54 Ibid., xi.
55 Quoted in Patricia Gebhard *George Washington Smith*, 10. David Gebhard, "Founding Father: George Washington Smith," http://www.architect.com/Publish/GWS.html. For an excellent analysis of the context for the rise of the Andalusian version of the SCR, see Kevin Starr, ch. 9, "Anacapa and Arcadia: The Santa Barbara Heritage," and ch. 10, "Castles in Spain: the Santa Barbara Alternative," in *Material Dreams: Southern California Through the 1920s* (New York: Oxford University Press, 1992), 231–62. Starr establishes the fact that Santa Barbara became an Anglo elite stage set for historicizing genteel Spanish romance, thereby providing an alternative urban model to the razzmatazz of Los Angeles.
56 H. Vincent Moses and Catherine Whitmore, *Henry L. A. Jekel: Master Architect of Eastern Skyscrapers and the California Style, 1895–1950* (Riverside, CA: Inlandia Press, 2017), op. cit.
57 Jekel's romanticized version of Arts and Crafts architecture carried over to his other period revival styles. "The Harbor," Jekel's 1931 French Norman Vernacular masterpiece, at Houghton and Pine in Riverside, readily demonstrated the similar rendering of his SCR and related Romantic Revival buildings.
58 California Arts and Crafts architects of the SCR, wishing to remain true to their materials, often chose not to paint the stucco coats of their houses and commercial buildings. Color preferences of later owners often led to painting over the natural tone, in favor of a "more" Spanish look of whitewashed walls.
59 Klotz and Hall, *Bungalows Revisited*, 275.
60 Ibid., 273–79, for a review of Spurgeon and his SCR houses.
61 For an overview of Wilson's work, see Klotz and Hall, *Adobes Revisited*, 312–16. An example of local press coverage of G. Stanley Wilson abounds in the *Riverside Daily Press*. See, for instance, *Riverside Daily Press*, Thursday, January 1, 1931, 15. The story includes Wilson's drawing of the north elevation of the Mission Inn with the new International Wing depicted at the northeast corner of 6th and Main Street.
62 Author's description from onsite observation, April 2016.
63 Robert M. and Kathreen Sanborn McCulloch Residence, "Bermejal," 1740 Canyon Road, 1924, 2011 Heritage Recipient, Redlands Area Historical Society, Inc.
64 Hugh and Aimee Tucker McCulloch Residence, 1766 Canyon Road, 1926, 2011 Heritage Award Recipient, The Redlands Area Historical Society, Inc. The Frank S. Mitten Residence, 1321 La Arriba Drive, 1929, is also designed as a high-end courtyard house, built on the crest of a steeply sloping hillside. This estate maintains a very high level of integrity, with most of its original SCR features intact. In turn, Redlands has middle and rarer working class iterations of the style at Glenwood, Clay, and Parkwood Drives in the old downtown.

Public structures in the SCR further adorn the neighborhoods of Redlands, including the historic First Baptist Church at 51 West Olive Drive and the old Fox West Coast Theatre at 123 Cajon Street. The Mission San Gabriel Assistencia, at 26930 Barton Road, constitutes a fanciful reconstruction of the original California Mission outpost. In 1937, the facility was moved a mile from its first location and rebuilt as a WPA project in the mode of the late 1920s SCR.

65 Author's description from onsite observation, April 2016.

66 Corona Historical Society Heritage Awards documentation for these houses is available in the Heritage Room of the Corona Public Library, along with the Historic Resources Survey Inventory records for the City of Corona, including these two houses.

67 For the best review of the SCR in Palm Springs and La Quinta, see Patrick McGrew, *Desert Spanish: The Early Architecture of Palm Springs* (Palm Springs Preservation Foundation 2012), op. cit. See also, by the Foundation, *Spanish Colonial Revival: Architectural Romance Comes to the Desert* (2012), for the Spanish Heritage Weekend event hosted by the Foundation and the City of Palm Springs. For the historic O'Donnell House, see Anthony A. Merchell and Tracy Conrad, *Ojo Del Desierto: The Thomas O'Donnell House Palm Springs* (The Willows Historic Palm Springs Inn: 2009), op. cit.; and see McGrew, *Desert Spanish*, 37–56, for the remainder of the Palm Springs residential structures. A few of the noteworthy designs of the region's signature architects include The El Mirador Hotel by Walker and Eisen, ca. 1926–27; Desert Inn (W. C. Tanner, ca. 1922–27); San Jacinto Hotel (Clark and Frey, ca. 1935: Movie Colony Hotel); La Quinta Hotel, ca. 1926, by Gordon Kaufman; Paul R. Williams, Casa Palmeras Hotel and Apartments, ca. 1930; Cody Hotel and Apartments, 1930, by Myron Hunt; and residential examples, such as Wallace Neff's "Ranchoa," the Arthur Keeler Bourne Estate, 1933, among many other.

68 Modern Architecture is defined as follows in the *McGraw-Hill Dictionary of Architecture*: "A loose term applied since the late 19th century to buildings in a variety of styles, in which emphasis is placed on functionalism, rationalism, and current methods of construction, in contrast with architectural styles based on historical precedents and traditional methods of building. This category often includes *Art Deco*, *Art Moderne*, *Bauhaus*, *Contemporary style*, *International style*, *Organic architecture*," *McGraw-Hill Dictionary of Architecture and Construction* (McGraw-Hill Companies, Inc., 2003). For an interesting example of twenty-first-century architectural mythmaking around Mediterranean ambience, see *Tuscan Style* magazine published by Special Interest Publications, Meredith Corporation, Des Moines, Iowa. See especially the Fall/Winter 2015 issue, which says that "Timeworn and timeless, these homes allow their owners to live surrounded by the comforts of the past while graciously moving into the future," 3. For a taste of Tuscan Style mythmaking, see articles in the Fall/Winter issue including "Lasting Heritage," "Modern Adaptation," "Mediterranean Panache," and "Continental Comforts."

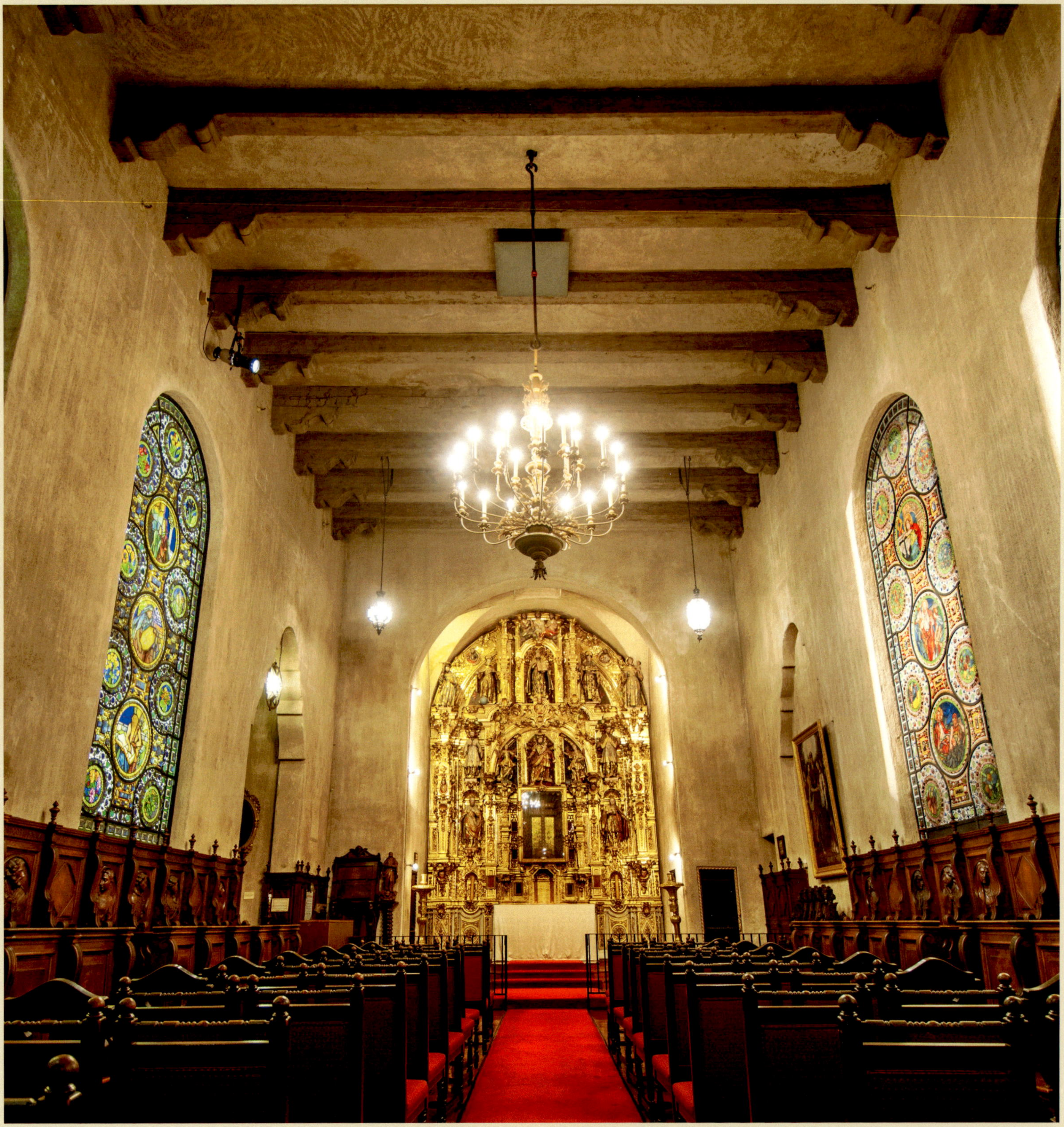

3

Lindsey Rossi

History by Design

Staging the Spanish Colonial Revival in the Inland Empire

Señora Moreno's house was one of the best specimens to be found in California of the representative house of the half barbaric, half elegant, wholly generous and free-handed life led there by Mexican men and women of degree in the early part of this century, under the rule of the Spanish and Mexican viceroys, when the laws of the Indies were still the law of the land, and its old name, "New Spain," was an ever-present link and stimulus to the warmest memories and deepest patriotisms of its people.

—*Ramona, A Story* (1884)

Warm and healing desert sands to the east; cool and refreshing coastal breezes to the west; fragrant citrus, roses, and vibrant bougainvillea abound in descriptions of Southern California from the early twentieth century. These visions of a blithe existence were rooted in romanticized notions of Spanish Colonial history like those described in Helen Hunt Jackson's popular novel *Ramona*, which, along with a surge in boosterism and the eventual popularity of automobile travel, propelled the influx of tourism and migration from the eastern states. The Inland Empire was often the first glimpse of California for passengers of the Santa Fe and Union Pacific railroads as they made their way to San Diego, Los Angeles, and points north.[1] With the export of citrus and import of tourists and settlers, the owners of irrigated orchards in such enclaves as Riverside, Redlands, and Highland propelled the Inland

Empire to unprecedented prosperity. In 1893, with the advent of refrigerated railroad citrus cargo transport, Riverside, California, boasted the highest per-capita income in the country.[2] While the Inland Empire served as the gateway to Southern California, local businessmen made it their duty to entice industry, tourism, and settlement by promoting the region's relationship with a past they did not hesitate to embellish.

With few exceptions, the essential contributions of architects and designers in the Inland Empire to the vanguard of the Spanish Colonial Revival have been largely overlooked. The proprietor and architects of Riverside's famous Mission Inn, with its eclectic design and collections, were initially inspired by the remains of California's many missions. This grand hotel set a thematic design precedent for the creation of neighboring structures and buildings throughout the Inland Empire. While the Mission Inn and the nearby Churrigueresque First Congregational Church predate the 1915 Panama-California Exposition in San Diego's Balboa Park, historians regularly overlook these regional contributions when attributing credit for igniting the craze for this ubiquitous architectural style in California. While the SCR flourished for nearly four decades among all socioeconomic classes, as the economic importance of the Inland Empire has waned, its contributions to the style's development and hegemony have been glossed over or simply forgotten. Extensive research has unveiled the role of the style in the overall design and interiors of key historic buildings, including the Mission Inn, the former home of the Riverside YWCA, the First Congregational Church, and two railroad stations, the Union Pacific's Kelso Depot, which served the Mojave Desert, and San Bernardino's Santa Fe Depot.

Notions of the west were significantly colored by literature, as fictionalized history proved more appealing than the harsh realities of frontier expansion. Jackson's enormously popular novel *Ramona: A Story* described Southern California after the Mexican-American War, during the 1870s and 1880s, as a sentimental paradise inhabited by pious Spanish Mexican ladies overseeing the happenings of their bucolic adobe homes amid fecund citrus trees. Her star-crossed lovers—Ramona, an orphaned, mixed-race, Scots-Native American girl, raised by an aristocratic family of Spanish descent, and Alessandro, a Native American shepherd—headlined a cast of respectful *caballeros* and demure *señoritas*. Jackson's evocations of the dignified traditions of Californio life suggested that the vital work of sustaining a typical rancho was conducted with serenity and ease by the early Californians.[3] The actual objective of her enchanting descriptions of daily life was to bring attention to the racism and cultural injustices endured by Native Americans, in the same way that Harriet Beecher Stowe's *Uncle Tom's Cabin* (1852) brought attention to the blight of slavery.[4] For most readers, however, the tragic love story overshadowed the social commentary Jackson intended. The novel quickly became a sensation,[5] staged repeatedly in colorful live productions, a tradition that persists in a desert town in Riverside County; since 1923, the *Ramona Outdoor Play* has been presented annually at the Ramona Bowl, an amphitheater in Hemet, California that makes use of the imposing natural landscape as the domain of the Indians, and features an adobe and tile-roof rancho façade that represents Ramona's quaint home. This setting evokes a sense of authenticity of place with its treacherous terrain scaled with ease by the Native American characters and the convivial celebrations of life on the rancho,

FIGURE 1 Postcard from the fifth annual *Ramona* outdoor play in Hemet, 1928

festooned with bright pink bougainvillea and Californios cavorting merrily in their colorful, traditional Spanish-Mexican dress (fig. 1).[6][7]

Jackson's story was so ingrained in the popular consciousness during Southern California's development that Ramona's name, along with Alessandro's, can be found in such varied locales as the names of towns and neighborhoods, schools, and even roads.[8] California State Route 2 was known as the Alessandro Freeway, while the 10 was called the Ramona Freeway. The two roads ran perpendicular with the 2 stopping short, never to meet the 10 in the end, much like the fictional lives of their namesakes. *Ramona*-inspired imagery spread far and wide as businesses of all types were named after the novel's characters and Ramona imagery was utilized in logos. The artistic creations of colorful citrus-crate labels include many portraits of Ramona, often as a maiden in colorful Spanish garb, with dark flowing hair, an idealized vision of the Spanish-Mexican *señorita*.

As tourists explored California, they frequently went in search of the locations described in the novel. A handful of sites became oft-visited tourist destinations due to similarities shared with the book's elaborate descriptions[9] including the del Valle family homestead in Ventura County, which legend made the iconic "Home of Ramona." An image of their old adobe home at Rancho Camulos—quaintly spare, surrounded by orchards, and decidedly simple compared to other labels with much more tenuous connections to the story—adorned the del Valles' own orange crates.

The fact that the Ramona story was pure fiction, and the locations described were most likely syntheses of places visited by Jackson when she traveled to California between 1881 and 1885, had little or no consequence to tourists infatuated with the tale. The search for evidence of this romanticized version of historic California drove the unabashed perpetuation of the Ramona myth. Entrepreneurs, eager to profit from the influx of tourism, used their financial resources and civic influence to bring to life fantastical designs, commonly rooted in splendid classical European and Spanish Colonial traditions, with a tenuous grasp—or outright denial—of the actual rough, rustic, and not-at-all stately Mexican California from the Spanish Colonial era. California missions were the oldest and largest expanses of architecture through the first half of the nineteenth century, but were derelict, often in ruins, and were never palatial. In fact, Spanish friars lived with few comforts and thus the missions were neither grand nor ornate beyond basic religious iconography and decorative painting (often completed by the indigenous converts). Photography from the mid-to-late nineteenth century shows low profile, unadorned adobes, false-front buildings, and shacks throughout much of Southern California.

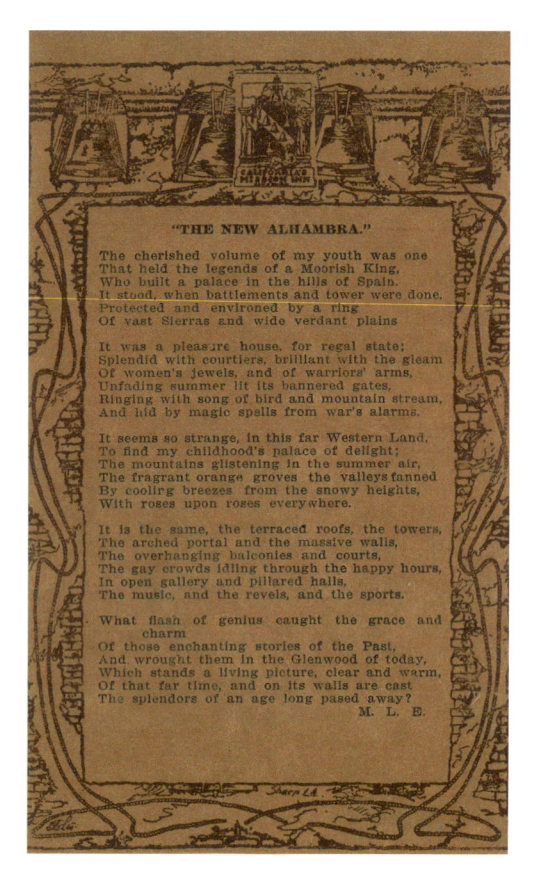

FIGURE 2 Postcard from the former Glenwood Inn featuring the poem "The New Alhambra" by M. L. E., ca. 1898–1915

Hotelier Frank Augustus Miller (1858–1935) of Riverside was one such impresario who founded his business on a mythology of early California that he helped to perpetuate (fig. 2).

Constructing the Myth

Miller's Mission Inn began with his father, Christopher Columbus (C. C.) Miller, as a two-story adobe rooming house, the Glenwood Cottage, in 1876, with a wooden addition constructed on the north side of the adobe two years later. In early 1880, C. C. Miller sold the cottage to his eldest son, Frank. By 1902, after garnering a reputation as a first-rate hotel man with a vested interest in creating a destination city, Frank Miller raised the funds to build a new hotel on the property; this would become the Mission Inn. At Miller's request, architect Arthur B. Benton gradually incorporated such dramatic flourishes as belfries, arched colonnades, and flared dormers throughout the hotel's construction—all reminiscent of the ecclesiastical design of California missions and Spanish Colonial architecture.[10] Prior to the 1903 grand reopening, when it was still called the Glenwood, Miller advertised the hotel as "a long, low, cloistered building of the Mission type, inclosing [sic] a spacious court and surrounded by magnificent old trees and palms. In the court the old adobe or casino adjoins the stately campanile with its sweet chime of old Mission bells."[11]

The hotel became a spectacular pastiche of Mission-inspired elements in the form of a bell tower resembling one at the sub-mission to Mission San Luis Rey de Francia, the San Antonio de Pala Asistencia (which had purportedly been visited by Padre Junípero Serra, founder of the first nine Spanish missions in Alta California); a dome modeled after Mission San Juan Capistrano; and buttresses like those at the Mission San Gabriel Arcángel in Los Angeles.[12] Likewise, the alternating hues of red brickwork of the Alhambra Suite are imaginatively reminiscent of the cathedral-mosque in Cordoba, Spain, and the small exterior lobed colonnade evokes the stylized arches of the Alhambra at Granada (fig. 3). Decorative elements such as tiles were frequently designed and created locally, and the vast assembled crew of skilled craftsmen that worked for the architects, crafted decorative structural elements such as colorful brickwork, *horror vacui* in the style of Churriguera, and whimsically eclectic turrets and domes (fig. 4).[13] The roofing tiles were described by E. F. Goff, Reverend of the First Congregational Church of Riverside from 1896 to 1907,[14] as covered with

FIGURE 3 Alhambra Suite and Fountain, Mission Inn, Riverside

FIGURE 4 Northeastern view of the Spanish Patio, Mission Inn, Riverside

"mosses and lichens and discolored with age" and claimed that they were made by the Indians under the guidance of Padre Antonio Peyri, the venerated predecessor of Junípero Serra, for the Mission San Antonio de Pala Asistencia of San Diego county, twelve miles from Temecula, in Riverside County. Whether by accident or by design, the reverend's words, like many of Frank Miller's, blur the line between the facts of the Inn's history and the true history of the region. He states the origins of the materials in a tone that is quite declarative, connecting the Inn's roofing tiles to an historic place, important historical figures, and points in time,[15] but due to a lack of definitive documentation, one cannot be sure whether the roofing tiles truly came from the Pala Mission or were based on the tiles from there.

Due to the increasing popularity and financial success of Miller's Mission Inn, he and his wife spent six months in 1911 traveling throughout Europe, six weeks of which were in Spain, purchasing antiques in each of the major cities and historic centers of traditional Spanish craft including Madrid, Toledo, Seville, Cordoba, and Granada.[16] Following Miller's return to Riverside, numerous crates filled with hundreds of pieces of Spanish art and artifacts were delivered to the hotel. Miller enlisted the esteemed Los Angeles architect Myron C. Hunt to design the Spanish Wing, built between 1913 and 1915.[17] In order to emphasize the historic Spanish tone of the hotel, Hunt and Miller incorporated these acquisitions wherever possible. Wood, purportedly from the Alhambra in Granada, was added in the construction of the south wall of the outdoor Spanish Patio. Likewise, Spanish wrought iron, escutcheons, and polychrome tiles were embedded into patio walls and fountains (figs. 5–6). Overlooking the Spanish Patio is the 1709 gilt-faced Anton Clock from Nuremberg, Germany. In 1953, statues of well-known historical California characters and symbols

FIGURE 5 View of Alhambra Suite balconies with wrought iron and polychrome tiles, Mission Inn, Riverside

FIGURE 6 Polychrome fountain tile detail, Mission Inn, Riverside

FIGURE 7 Anton Clock in the northeast corner of the Spanish Patio, with the automaton figure of Juan Bautista de Anza, Mission Inn, Riverside

FIGURE 8 Postcard of the Sherman Institute, United States Indian School, Riverside, ca. 1905

were installed below the clock, mechanically rotating on the quarter hour (fig. 7). The figures include Padre Junípero Serra (1713–84); Juan Bautista de Anza (1736–88), who proposed to the Viceroy of New Spain expeditions throughout the present-day southwestern United States and California; Saint Francis of Assisi (ca. 1181–1226), whose teachings and practices were the foundations of the Franciscan Order of Catholicism, which established the California missions and something of a hotel mascot; a bear, the enduring flag symbol for the republic and subsequent state of California; and a Native American representing the indigenous inhabitants and converted subjects under Spanish colonialism in California as well as a reference to the significant Indian population in the region, many of whom were educated at the nearby Sherman Institute, which also became a noted tourist stop (fig. 8). Seemingly benign references to the region's Indian population and the government school building as another beautiful representation of the SCR belie the hard facts of its purpose. Young Native American students of the Sherman Institute were stripped of their language, families, and culture, conditioned to be "civilized" according to middle class Anglo values, and trained to be subservient to that class. Given this, coupled with the Anglo re-interpretation of Spanish Colonial and Mexican history as seen in SCR architecture and design, one may understand the use of the SCR at the institute to have been part of this "civilizing" process.

Upon first glance, the Anton Clock's collection of characters may seem like a simple medley of quaint figures; however, along with the extensive collection of Spanish and Mexican artwork installed throughout the Spanish Patio, this is one of many instances in which Miller intermingled Spanish and Mission-inspired icons to communicate the story of California's Spanish colonial history within his hotel, thus insinuating that his Mission Inn was at one time an integral part of it, however nebulously.[18] By disregarding and repositioning the facts of the Spanish Colonial Mexican and Native American material culture that composed his inn, Miller—and other promoters of the style genre—was effectively participating in the deracination of California's Spanish Colonial past in favor of a mythology that, despite flights of international whimsy, was largely Eurocentric. Miller catered almost exclusively to an elite Anglo clientele that had not yet reached a popular moral obligation to question what he presented as anything other than fact and beauty for beauty's sake.

The interior construction of the Mission Inn was predominantly Mission Revival with eclectic international decorative flourishes creating inviting places to sit and absorb the unique exotic feast for the eyes.[19] Broad, square, non-tapered Craftsman-style pillars were set down the length of the lobby and capped with minimal molding. Imposing oak beams spanned the width of the ceiling creating a space that was at once grand, relaxed, and inviting (fig. 9).

Though Miller was not educated past elementary school, his wife, a schoolteacher, taught him about art and history, and once Miller's zeal for collecting was piqued, it never ceased. Miller and his wife's travels extended to Japan, China, Mexico and Europe from which they shipped home to Riverside more objet d'art, and objet de curiosité. The sums the Millers spent on these acquisitions became a draw nearly as powerful for tourists as the Mission Inn that displayed them.

True to Miller's fantastical vision of Spanish Colonial tradition, and in keeping with the late nineteenth-century propensity for the

FIGURE 9 Lobby of the Mission Inn ca. 1940

"cluttered look," his furniture collections were not displayed with strong regard for the manner in which a proper Spanish home would have been arranged.[20] This, despite Miller's proclaimed studies of the Spanish interiors he visited during his travels in Europe.[21] Other than in the Spanish Gallery, in which it seems Miller most closely adhered to Spanish decorative traditions, he arranged furniture, even if it was historical or intended to look as if it was, in decidedly late-nineteenth-century conversational configurations.[22] Disregarding the highly angular structure of Mission style and traditional Spanish furniture, which was intended to be arranged either in linear configurations or stored up against walls, respectively, Miller brought the lobby furniture out from the periphery and forced it into angular, tête-à-tête, and group orientations.[23]

Fine porcelain, Oriental rugs, ornate Mexican, Spanish, Italian, and early modern paintings and furniture, religious iconography, and Native American artifacts added layers of colorful, historically inspired splendor to the overall weighty oak ambiance of the Inn. Though the hotel had become famous for its beguilingly eccentric and ever-expanding collections, Miller sought to present the Mission Inn as a legitimate center for important art exhibitions and used his new Spanish Art Gallery as the setting for two prominent exhibitions that showcased the Mission Inn's collections along with important pieces on loan from the renowned Ehrich Galleries in New York, of which he was a patron. For the New Year's Eve 1914 grand opening of the Spanish Art Gallery, Frank Miller hosted an exhibition of Spanish and Mexican paintings from the Inn's collection along with the works *Old Darby* (1884) and *Roman Warriors* (nineteenth century) by French artist Rosa Bonheur (1822–99)[24], one of the nineteenth century's most esteemed woman artists. Likewise, Riverside architectural artist William Charles Tanner (1876–1960), who specialized in historic early California scenes, loaned pieces for the exhibition.[25] Miller developed a particularly keen interest in paintings which he acquired during his travels to San Francisco, New York, and abroad. During a 1916 trip to New York's Ehrich Galleries, Miller initiated a 1916 collaborative exhibition titled *Old Spanish Masters*, which featured loaned works by the revered Alonso Sánchez Coello, El Greco, Bartolomé Esteban Murillo, Jusepe de Ribera, and Francisco de Zurbarán.[26] He admired work by landscape painter William Keith (1838–1911), initially purchasing the artist's *Sunset Glow* and, in 1918, the sweeping panorama *California Alps*, which Miller first viewed during his visit to the 1915 Panama-Pacific Exposition in San Francisco. These displays epitomized the ancient beauty and grandeur of the West, which Miller sought to present to his visitors at the Inn in an effort to assert the region's own regal past while also bolstering Miller's efforts to spuriously place his Mission Inn within the ranks of the vital Spanish Colonial history of California.

Enticing Them to the Show

Curiosities and decorative objects were gradually added to the collection by Miller as well as his trusted curator of artifacts, Dr. Francis S. Borton (1862–1929). While Borton's role is not explicitly documented with regard to the acquisition of certain Mexican pieces, he is largely credited with acquiring many of the Mexican objects that came to the Mission Inn between 1910 and 1929. His ties to Mexico City were strong owing to his twenty-year tenure there as an archaeologist and Methodist missionary from 1890 to 1910. Barton spoke fluent Spanish and was well acquainted with the Mexican

FIGURES 10 & 11 Della Robbia bas-reliefs depicting the Annunciation (20th c. Italian copy of a 1522 original), mounted into the Spanish Patio's eastern wall, Mission Inn, Riverside

market for art and artifacts. In a similar manner as William Randolph Hearst's future Hearst Castle (begun in 1919 and designed by Julia Morgan), Miller and his architects collaborated to incorporate into the structure and decor of his Mission Inn architectural fragments and treasures procured during his international travels. Sixteenth-century carved wooden doors from Granada lead into the Spanish Art Gallery. Iron balcony grills were hand-wrought in Spain and decoratively carved columns originated from Mexico. Tiles from both California and the Triana tile factory in Seville, along with the crown from the Cordoba municipal fountain were incorporated into the gargoyle fountain on the Spanish Patio. Cloister pews were brought to the Mission Inn and appropriated for mostly secular display.[27] Two bas-reliefs replicas by Della Robbia representing the Annunciation are incorporated into the north wall of the Spanish Patio (figs. 10–11). Decorative objects and curiosities placed throughout the Inn included Spanish cannons; European heraldry; free-standing sculptures; bells from all over the world including Spain, Italy, Japan, India, Mexico, and California churches; crosses and candelabra; wax figures of Pope Pius X (1858–1914; elected Pope 1903) and thirteen attendants; and folding X-framed (or Savonarola) chairs (figs. 12–13).

Guestrooms had clean white walls and arched windows intended to resemble simple—if relatively plush—monastery quarters (fig. 14).[28] The original writing desks and chairs in many of the guestrooms were designer Gustav Stickley's famous Mission-style creations, custom-made for the Mission Inn.[29] Abundant Mission revival furniture by Stickley (or made in his style) and Charles Limbert, as well as Colonial-style and Colonial Dutch-style tables and chairs, and rocking chairs, were positioned throughout the hotel.

FIGURES 12 & 13 Postcards of the Cloister Music Room of the Glenwood Mission Inn, 1909

Frank Miller playfully and shrewdly paid homage to the Franciscan friars and the California missions, endeavoring to place the Mission Inn within the accepted history of the California missions, even though no missions were ever constructed in Riverside County.[30] Virtually every corner of the Inn presented an *assumed* splendid creation of the bygone Spanish Colonial era—one expressed by world travel, religiosity, and sentimentalized views of regal colonialism and quaint indigenous cultures. The result was a grand spectacle of intentional romanticization and obfuscation. Perhaps the most extravagant display of Miller's intent to draw visitors via implied connections to the faux religious history of the Inn can be seen in the Saint Francis Chapel. The exterior entrance is adorned in the Churrigueresque style with a Tiffany rose window depicting the seasons and inscribed "Hast Thou Made Them All" installed above wooden doors (figs. 15–16). The nave of the narrow marriage chapel is flanked on each side by three Tiffany stained glass windows, symbolizing the months of the year with floral symbols of each season and the names of the months inside medallions (fig. 17). The chapel was designed specifically to display these windows along with the eighteenth century Spanish Colonial baroque, gilded, carved-wood altarpiece, or reredos, dismantled into pieces and shipped in 1921 from the private chapel of a former silver-mining family in Guanajuato, Mexico, who had lost their fortune.[31] The altar initially displayed the Holy Family surrounded by nearly-life-sized figures and surmounted by the Holy Trinity, however, the family chose to keep the figure of the Virgin (fig. 18).[32] Though not officially consecrated by the Catholic Church, the St. Francis chapel nonetheless served as—and continues to be—a popular location for weddings.

Like so many entrepreneurs of the time, Miller advertised a diverse roster of curiosities worthy of visiting and capitalized on the popular new vacation trends: car travel and emblematic landmarks. *Touring Topics* (1909, renamed *Westways* in 1934) was an early publication produced by the Automobile Club of Southern California that advertised routes and points of interest for travelers interested in the novel and increasingly fashionable road trip vacation. California's missions were featured prominently in articles and along routes recommended by the Auto Club. In 1906, the first bell commemorating El Camino Real was unveiled in Los Angeles, and 449 bells were subsequently installed from San Diego County north to Paso Robles and eventually as far north as Sonoma, and were cared for by the Automobile Club from the mid-1920s to the early 1930s.[33] The bells were placed along El Camino Real, the historic 600-mile route connecting all

FIGURE 14   Photograph of Mission Wing guest room at the Mission Inn, 1903

FIGURE 15   Rose window by Louis Comfort Tiffany set above the Saint Francis of Assisi Chapel doors, Mission Inn, Riverside

FIGURE 16   Detail of the Winter section of the Tiffany rose window, which depicts the four seasons, Saint Francis of Assisi Chapel, Mission Inn, Riverside

FIGURE 17   Interior of the Saint Francis of Assisi Chapel with light fixtures designed by Frank Miller and G. Stanley Wilson, the gilded Rayas Altar, Tiffany stained glass windows, and wooden choir stalls designed for the chapel with carved reliefs of the heads of saints originally from a Belgian convent

FIGURE 18   Rayas Altar in the Saint Francis of Assisi Chapel, Mission Inn, Riverside

89

twenty of Alta California's Spanish missions as well as four presidios, three pueblos, and several sub-missions.[34] Travelers used the bells (hung from a stylized shepherd's hook), as reference points to explore the old routes of the Franciscan friars. The bells, with signs indicating mileage and location, were so relied upon as beacons of safe, manageable routes that they became something of a highlight of their own for the motorist. One itinerary, from a 1915 issue of *Touring Topics*, describes a route from Los Angeles to San Diego's Panama-Pacific Exposition, which opened that year. The illustrated map reads:

> Take the Mission Road to San Diego from the Little Mission Church on the Los Angeles Plaza, visiting San Gabriel Mission and passing through the Mission orange groves and vineyards to Riverside. See the great Government Indian School and then ascend Mt. Roubidoux to Father Serra's cross from the sunset and spend the night at the Mission Inn. Continue through the Ramona country, close to the high mountains, past Lake Elsinore's olive groves and seeing the three Missions of Pala, San Luis Rey, and San Diego. The journey ends at the Exposition City ....[35]

The Mission Inn is included within the roster of missions recommended to the tourist, and, without any other explanation, likely mislead motorists into assuming the Inn was a contemporary of the missions.

While it was unlikely that anyone would assume that the El Camino Real bells were in fact from the mission era, Frank Miller was well acquainted with the assumptions simple objects such as bells could invoke from a visitor and, again, utilized this practice of blending the imaginary with the genuine to great affect. Though his international collection of bells as well as crosses was considered to be one of the largest in the western world, one in particular stood out from the rest. Known as the Raincross, the design was dreamed up by Miller and Arthur Benton between 1907 and 1909 for use as a street light, purportedly taking inspiration from both a bell and a cross from Miller's immense collection.[36] Though there are a variety of unverifiable origin stories for the bell and cross design, the most plausible seems to be that the bell and cross composite was inspired by a Navajo double cross used in prayers for rain and a replica of a bell ostensibly used by Padre Serra. With the superficial resemblance of the bell to something ecclesiastical, the Raincross implies origins rooted in the early California Missions. Originally the emblem of the Mission Inn, the Raincross symbol was subsequently adopted by the city, with permission from Miller (fig. 19). Most famously, the Raincross was applied to the top of lamp poles throughout the downtown area, with bell-shaped lights hanging from the bottom of the frame. Like virtually everything that Frank Miller did with regard to local boosterism, this utterly misleading icon nonetheless

FIGURE 19   Raincross symbol, created by Frank Miller and Arthur Benton, incorporated by the city of Riverside into street lamps beginning in 1909

became the ubiquitous symbol of Riverside, demonstrating how Miller successfully sold his mythology of Riverside not just to his guests, but to the city itself.

A Matter of Civic Pride

As Miller commissioned the ever-expanding Mission Inn as a virtual SCR wonderland, he and other civic leaders worked to promote a consistent aesthetic reflected in the buildings built throughout the 1920s in variations of the style with details created by renowned architects and artists. In 1929, Julia Morgan, architect of the famed Hearst Castle, and the first licensed woman architect in California, was commissioned by the Young Women's Christian Association to design their new facility in Riverside, now home of the Riverside Art Museum.[37] Morgan designed over thirty buildings for the YWCA in the United States, including several throughout California, five of which were in Southern California. Recommended to the YWCA by Phoebe Apperson Hearst, a patroness of Morgan and mother of William Randolph Hearst, the architect worked with the organization to create ideal spaces where women could enrich their lives, find a sense of community, and provide them with safe shelter.

As with all of her clients, Morgan worked closely with the YWCA to design facilities that were tailored very specifically to their needs. She often adhered to a basic floor plan that the organization found to be ideal for the programs provided at most locations, and the design and decoration frequently included a courtyard or central atrium, archways, colonnades, small balconies, and decorative wrought iron. With the exception of the Craftsman-style Harbor Area YWCA in San Pedro (1918), Julia Morgan's YWCAs of Southern California tend to be highly reflective of her penchant for Mission Revival and Italian Renaissance Revival architecture, and yet each exhibits its own character. Morgan designed YWCA buildings for San Pedro (Harbor Area, 1918); Pasadena (1921); Long Beach (1923); Hollywood (Hollywood Studio Club, 1925); and Riverside (1929).[38]

Julia Morgan's Riverside YWCA was a substantial departure from the sumptuous take on the revival styles preferred by the influential businessman and Mission Inn owner, Frank Miller. The comparatively modern façade and concrete structure of the neighboring Riverside YWCA was not in keeping with Miller's vision for an aesthetically unified downtown area with his extremely historicized luxury hotel at the center of all activities. Despite Miller's vocal misgivings about Morgan and her designs, the YWCA would not be dissuaded from their choice of architect, who had a proven understanding of their institutional needs. Furthermore, the women of the YWCA preferred the gently hybridized, feminine aesthetic proposed by Morgan.

The interiors of the Riverside building are composed in a relatively simple collection of corridors and square rooms surrounding a bright and airy central atrium, creating a virtually blank canvas upon which to include an elegant array of stylistic details. Original blueprints illustrate a simple layout wherein Morgan cleverly achieved depth and variation using small stairwells, balconies, colonnades, and columned windows and doorways. Column and arch structures and extensive wrought-iron work throughout the museum are highly evocative of the ubiquitous porticos of California missions and Mediterranean revival buildings. Delicate, detailed interior decoration, such as hand-painted fireplace tiles and the sculptural floral motif on the column capitals, both designed by Morgan and created by her craftsmen, can be perceived as flourishes of Spanish, Hispano-Moresque, and Arts and Crafts styles (fig. 20).[39] She also chose to expose the broad concrete beams in the lobby ceiling perhaps as a nod to Arts and Crafts and SCR architecture.

The Riverside YWCA building has undergone numerous renovations since its construction. An indoor pool featuring pink and green tiles, the official colors of the YWCA, was filled in to create the current Art Alliance Gallery.[40] The removal of a balcony that originally overlooked the pool's perimeter opened the space to accommodate large art displays. Dormitories and pool dressing areas became essential museum storage, while the gymnasium is now one of the museum's largest galleries. A stage at the far end of the room is now discreetly concealed and contains the climate control systems. Renovations transformed the building's rooftop gaming area into a comfortable event space, complete with an outdoor fireplace. Original lighting fixtures remain and continue to illuminate the space as they have since 1929. The first floor office became the museum gift shop, while upper classrooms and meeting areas now contain offices, classrooms, and gallery spaces. A plaster Madonna and Child low-relief roundel, designed by Morgan and presented as a gift in 1929, had been incorporated into the wall above the fireplace, but was removed after the building changed hands in 1967 and its whereabouts are unknown.[41] The interior has since been painted with subtle shades intended to highlight the interior decorative elements, most noticeable on the columns in the lobby. Most of the metal fixtures such as the atrium door handles remain unchanged. Screens and glass obscured the second floor loggias for several decades, but the Riverside Art Museum has since removed them to showcase Julia Morgan's original design.

While the Mission Inn and YWCA featured the creative talents of its architects and craftsmen, the First Congregational Church, already considered a masterpiece for its influential Churrigueresque design, has been the recipient of several exceptional pieces of religious artwork. The first significant gifts were stained glass windows designed by Horace Judson and crafted by the historic Judson Studios in Los Angeles. The *Sistine Madonna* (1916) was the first to be installed with five more dedicated in 1947 and three more dedicated in 1955 (fig. 21).[42] In 1927, a hand-carved baptismal font by Swiss carver Richard Lippich of Lancaster, New York, was given to the church by congregants who, like so many in California, were transplants from the East Coast (fig. 22). The inscription on the font reads, "In memory of one from the east who worshiped in this church."[43] Perhaps most prominently displayed is the Great Cross in

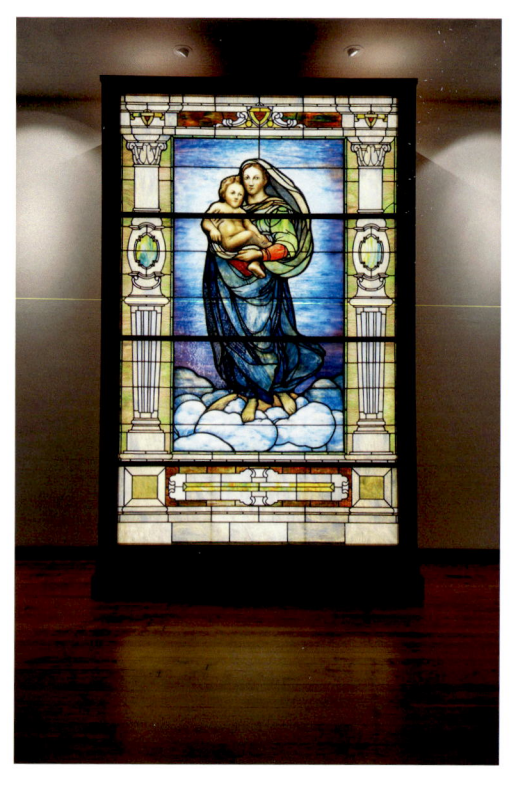

FIGURE 20 Polychrome fireplace tiles and carved oak mantel designed by Julia Morgan, in the lobby of the Riverside Art Museum (formerly the Riverside YWCA), built in 1929

FIGURE 21 Sistine Madonna window designed by Horace Judson, made by the Judson Studios of Los Angeles, for the First Congregational Church, Riverside, in 1916

FIGURE 22 Baptismal Font carved by Richard Lippich of New York, gifted to the First Congregational Church, Riverside, in 1927

FIGURE 23 The Great Cross designed by Margaret Montgomery and conceived of by Millard Sheets, for the First Congregational Church, Riverside, in 1950

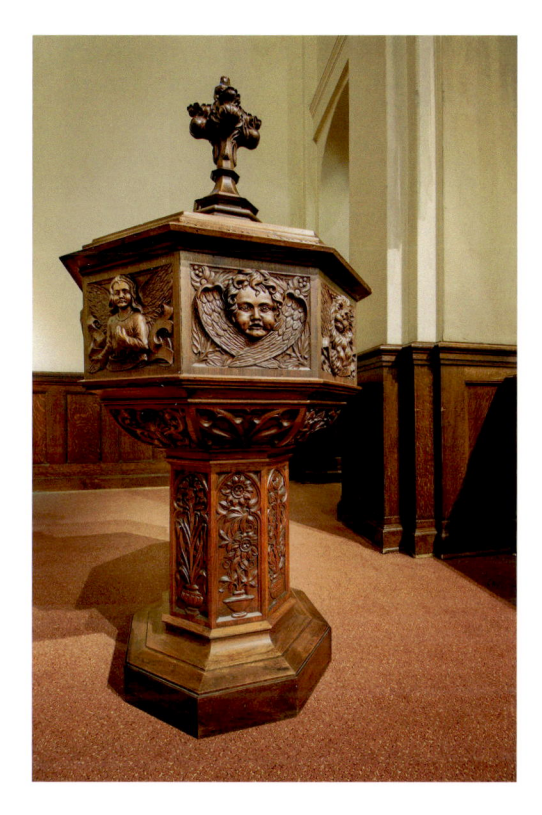

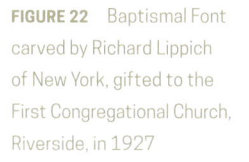

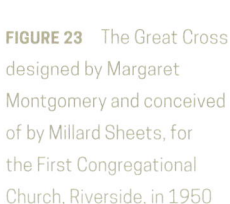

the Sanctuary. Originally conceived of by the world-renowned Millard Sheets (1907–89) and designed by enamel artist Margaret Barlow (née Montgomery), the cross is decorated with the central religious beliefs of the church on twenty-three enameled copper panels (fig. 23).[44]

Welcome to Romantic California

Before setting eyes on the Mission Inn or any of the region's other monumental structures, travelers from the east would have passed through a train station built by one of the two railroads that serviced Southern California. The Santa Fe Railroad built depots in Barstow and Riverside, as well as the station at San Bernardino that became its landmark in the Inland Empire. The Union Pacific Railroad, long interested in securing a stronghold in industrial and tourist transport to California, established its depot in the Mojave Desert, at Kelso, along the Salt Lake Route. Both train depots served as grand entrances into California and remain as ideal public spaces to explore the development and impact of the Spanish Colonial Style in the Inland Empire. They effectively set the stage for tourists looking to find the romantic place about which they had surely fantasized.

In 1886, a two-story wooden structure functioned as the San Bernardino train station and subsequently burned down in 1916. With the support of local civic leaders, the city hired architect W. A. Mohr to design and construct the future depot. The exterior of the Santa Fe Depot in San Bernardino, built in 1918, combines Mission Revival and Hispano-Moresque Revival styles.[45] The interior was distinctly Mission Revival, with walls in the main lobby wainscoted in red, green, and white tile to the height of the doors and floors covered in red-tile pavers. Tiled columns with composite capitals reached up to a coffered ceiling. Double-sided wooden benches in the waiting rooms were constructed in the unpretentious, yet weighty Mission style (fig. 24).[46] The exterior combines a front elevation in the classical order copied directly from the original Mission Santa Barbara, but presented with rounded elements in a contemporary awe-inspiring fantasy. Combined with the Mission-style sobriety of the interior, the heft of the overall design conveyed a sense of historical fact, declarative in its sense of place and time.

The Union Pacific depot in Kelso, located nearby a spring water source, an oasis deep in the arid wilderness of the Mojave Desert, initially provided the most basic food services for trains without dining cars.[47] However, in an effort to compete with the extremely popular Harvey House lodgings and eateries that followed the Santa Fe lines, Union Pacific developed their answer to the comfortable hotel-restaurant chain with the Kelso Depot &

FIGURE 24 Mission Revival interior of the Santa Fe Depot, San Bernardino, opened in 1918 and renovated in 2004

Clubhouse, built in 1923.[48] The exterior was definitively Mission Revival with a stucco façade, an arcade with buttressed corners and curved Mission-style parapets, and a terracotta tile roof. Inside, the depot contained a Craftsman-style staircase; plaster in an uncomplicated texture covered ceilings and walls above panel-and-batten wainscoting.[49] The floor of the arcade and interior was scored, red-tinted concrete, mimicking glazed terra cotta tile pavers. Facilities for travelers included a restaurant, baggage room, and telegraph office. For the staff and railroad crewmen were sleeping quarters, a billiard room, locker room, library, as well as a conductor's room.[50]

Although the Kelso depot was not as grandiose as the San Bernardino Santa Fe Depot, its understatedness nevertheless typified the air of "Spanish" California to many a weary traveler. Its remote location and emblematic design would have been a welcome respite and put travelers at ease amid their new environs, whereas the Santa Fe depot commanded the viewer into their new surroundings. These detailed Inland Empire train stations, hotels, churches, and civic centers presented a first glimpse at California's "Spanish" wonderland.

Design innovators and entrepreneurs were intricately immersed in the early creation of what became the widely accepted built history of Southern California. By catering to the fanciful whims of prominent clients, architects and designers reflected the creative means by which promoters were willing to draw visitors and migrants to their cities. Design eccentricities deeply rooted in the basic idea of Spanish Colonialism largely glossed over the facts that were far less alluring and, indeed, often harsh in their truth. Frank Miller indulged himself by constructing a machination that served both his personal obsession with collecting as well as his professional intent to place his hotel and city within an imagined version of history. Miller's grandiose narrative of a regal Spanish Colonial past included extreme decorative flights of fancy while also being anchored in the supposedly simple lives of the pious Spanish friars, Native Americans, and Mexicans who once lived on California's mission grounds. The false narratives portrayed in the SCR architecture and design of the region naturally led to a predilection to whitewash the regional history in an effort to draw tourism and a wave of new, elite, mostly Anglo residents. Likewise, railroad companies capitalized on the popular assumptions made by visitors viewing the built environment of Southern California for the first time. Thus, these train depots seemed to affirm these fantasies via archetypal exteriors that rose to meet the weary traveler and interiors that set them on a course to explore and supposedly understand their new surroundings.

Notes

1. Esther Klotz and Kevin Hallaran, "Citrus chronology," in *A History of Citrus in the Riverside Area*, 2nd edition, ed. Esther Klotz, Harry W. Lawton, and Joan H. Hall (San Bernardino, CA: Riverside Museum Press, 1989). In 1886 the Santa Fe railroad opened a direct line to Riverside, and in 1893 the first refrigerated rail cars took oranges east.
2. Ronald Tobey and Charles Wetherell, "The citrus industry and the revolution of corporate capitalism in Southern California, 1887–1944," *California History* 74 (1995): 6–21; or, see JSTOR 25177466 at www.jstor.org.
3. Dydia DeLyser, *Ramona Memories: Tourism and the Shaping of Southern California*. (Minneapolis: University of Minnesota Press, 2005), 196–97. See DeLyser's explanation referencing *Racial Fault Lines* by Tomas Almaguer and *Conquests and Historical Identities in California* by Lisbeth Haas. Californios are generally classified as having been of Spanish ancestry, born in present-day California or Baja California, from Portolá's arrival in 1769 to California's secession to the United States in 1848. The term was used to distinguish between the landed Spanish/Mexican gentry and the Native Americans or Mexicans who intermarried with Native Americans.
4. Jackson incorporated real events into the novel to bring undeniable truth to the story including the eviction of Luiseño Indians from Temecula and the confiscation of their cattle in 1875 by the Sheriff of San Diego County. Jackson's references to American takeovers of Mexican-granted lands deeded to Californios into the new public domain and unjust killings of indigenous peoples were all based on factual occurrences. Helen Hunt Jackson was praised, along with Harriet Beecher Stowe, as one of the best women authors of her time and for her literary efforts on behalf of Native Americans.
5. The story was first published as a weekly serial in the *Christian Union*. Over 600,000 copies of Ramona had been sold sixty years after it was published, and the book has never been out of print.
6. Don Shirley, "Ramona: Hit and Myth Affair: Not just a tourist attraction, the annual Ramona Pageant in Hemet also embodies one of the richest ironies of Southern California culture," *Los Angeles Times*, April 22, 1993, http://articles.latimes.com/1993-04-22/entertainment/ca-25980_1_ramona-pageant. The Ramona Pageant minimally referenced the social issues that concerned Jackson and focused much more on the love affair and travails of the main characters. It also presented rancho life as much simpler than the realities discussed in the novel.
7. State Outdoor Play, January 11, 1993.
8. Ramona High School, Riverside; town of Ramona, San Diego County, California; Alessandro Heights neighborhood, Riverside; Ramona and Alessandro roads in numerous Southern California cities.

9   DeLyser, "Rancho Camulos: Symbolic Heart of the Ramona Myth," in *Ramona Memories*, 65–84; DeLyser, "Ramona's Marriage Place, 'Which No Tourist Will Fail to Visit,'" in *Ramona Memories*, 99–119; Peter Y. Hong, "Pieces of State's History Escape Fires," *Los Angeles Times*, November 2, 2003. DeLyser describes the del Valle family's Rancho Camulos in Ventura County and the influx of tourists eager to see the fabled adobe "Home of Ramona." As a working ranch with a private homestead that greatly resembled the one described in the novel, life was regularly interrupted by unannounced tourists who were offered food and drink according to proper Californio tradition. The Union Pacific railroad even extended a tourist line to stop at the ranch and other sites made famous by the novel. Unfortunately, the costs of tending to guests, who were unaware of the cultural and financial demands they were making and who frequently had little respect for the fact that this was an actual family home, vandalized the property and stole odds and ends as mementos, picked citrus from the working orchards, and, all told, were a financial drain. On the other end of the spectrum, the Estudillo adobe, "Ramona's Marriage Place" in Old Town, San Diego, flourished as a designated pilgrimage stop for newlyweds and other Ramona fans. After a series of sales, San Diego industrialist John D. Spreckles purchased the building and restored it with a curio shop to purchase postcards and mass-produced trinkets imprinted with the image of the site or with Ramona's name.

10 Esther Klotz, *The Mission Inn: Its History and Artifacts* (Corona, CA: UBS Printing Group, 1993), 5–8.

11 Frank Miller, "The Glenwood," *Sunset Magazine*, 1903.

12 Klotz, *Mission Inn*, 19. Postcard, "The Monastery, The Mission Inn, Riverside, California."

13 University of California, Riverside, Special Collections Archive, Peter J. Weber Papers.

14 D, W. D. "The Pacific Slope: From Pastorate to College," *The Congregationalist and Christian Word*, June 15, 1907, 814.

15 Reverend E. F. Goff, "In the Orange Grove City: A Few Facts Concerning the Past and Present and the Unparalleled Progress of Riverside, California," *Sunset Magazine* (March 1903): 397, 402.

16 Klotz, *Mission Inn*, 21, 23, 28. Miller's first wife, Isabella, died in 1908. He married Marion Clark in 1910.

17 Inventories and photographs, Mission Inn Museum archives, Mission Inn Museum, Riverside, CA; Klotz, *Mission Inn*, 31–42. Painting, sculpture, decorative arts, and ephemera are far too numerous to list in this essay.

18 Inventories and photographs, Mission Inn archives; Klotz, *Mission Inn*, 195.

19 Ibid., 10, 33, 41.

20 Arthur Byne and Mildred Stapley, *Spanish Interiors and Furniture* (New York: Dover Publications, Inc., 1969), i–ix; Peter Thornton, "1870–1920," in *Authentic Décor: The Domestic Interior 1620–1920* (London: Seven Dials, 2000), 320. Byne's numerous photo plates illustrate Stapley's explanations of traditional Spanish interior configurations.

21 Maurice Hodgen, *Master of the Mission Inn: Frank A. Miller, A Life* (North Charleston, SC: Ashburton Publishing, 2014), 206, 217–18.

22 Inventories and photographs, Mission Inn Museum archives; Thornton, "1870–1920," op. cit., 344–45, 353, 360.

23 Goff, "In the Orange Grove City," 395, 400, 401; Thornton, "1870–1920," op. cit, 344–45, 353, 360. Published photo plates show the Mission Inn lobby, corridor alcoves, and guest rooms crowded with all manner of Mission-style furniture, writing desks, rocking chairs, plush wing chairs, Dutch chairs, expansive Persian rugs, landscape and historical paintings on the walls, and displays of Native American, Spanish, and Mexican pottery.

24 Peter J. Holliday. *American Arcadia: California and the Classical Tradition*. (New York: Oxford University Press, 2016), 395. Loaned from San Francisco; Klotz, *Mission Inn*, 37.

25 Edan Hughes, *Artists in California, 1786-1940* (Sacramento: Crocker Art Museum, 2014); Klotz, *Mission Inn*, 37-38. Tanner is also known for his murals in the kitchen of the Mission Inn.

26 Klotz, *Mission Inn*, 42.

27 Inventories and photographs, Mission Inn archives; Klotz, *Mission Inn*, 194-95.

28 Ibid., 13.

29 Ibid; David M. Cathers, introduction, "The Work of L. & J.G. Stickley" in *Stickley Craftsman Furniture Catalogues* (New York: Dover, 1979), iii–viii; Gustav Stickley, *The 1912 and 1915 Gustav Stickley Craftsman Furniture* Catalogs (Philadelphia: Athenaeum, 1991), 5-13. While not all guestroom furniture was genuine Stickley, a significant portion was, at least, extraordinarily similar.

30 San Bernardino County Museum, "The 'Asistencia' California Historical Landmark #42," http://www.sbcounty.gov/museum/branches/asist.htm. In 1819, San Bernardino de Sena Estancia, the ranch outpost of Mission San Gabriel Arcángel, was built in Redlands, San Bernardino County. What is now the city and county of Riverside was effectively surrounded by sub-missions and missions, including Mission San Juan Capistrano and Mission San Gabriel Arcángel to the west, Mission San Luis Rey de Francia and Mission San Antonio de Pala Asistencia to the southwest and southeast, respectively, and San Bernardino de Sena Estancia to the east.

31 Hodgen, *Master of the Mission Inn*, 239-41.

32 Ibid.

33 "FAQs." California Department of Transportation, 2016, http://www.dot.ca.gov/hq/paffairs/ faq/faq62.htm (accessed June 2016).

34 Sub-missions were self-sustaining satellite mission locations that did not have resident priests and, like other missions, were typically located near concentrations of Native American populations.

35 Kathy Talley-Jones and Letitia Burns O'Connor, *The Road Ahead: The Automobile Club of Southern California 1900-2000* (California: Automobile Club of Southern California), 37.

36 Klotz, *Mission Inn*, 17-18; *The Bells and Crosses of the Mission Inn and Ford Paintings of the California Missions*, Mission Inn, Riverside, CA; Emily Ann McEwan, "Southern California's Unique Museum-Hotel: Consuming the Past and Preserving Fantasy at Riverside's Mission Inn, 1903-2010," (PhD diss., UC Riverside, 2014). Miller was particularly fond of drawing associations between objects in his collections and Padre Junípero Serra. For a complete investigation into this phenomenon, consult McEwan's dissertation.

37 Riverside Public Library, Special Collections Archives, Riverside YWCA Papers; Riverside Art Museum, Archives and Permanent Collection.

38 Cal Poly University, San Luis Obispo, Special Collections Archives, Julia Morgan Papers; ibid., Sara Holmes Boutelle Papers.

39 "Hispano-Moresque," "A Dictionary of Architecture and Landscape Architecture." *Encyclopedia.com,* http://www.encyclopedia.com/education/dictionaries-thesauruses-pictures-and-press-releases/hispano-moresque (accessed December 4, 2016). The term is used to describe architectural styles in the Iberian Peninsula which incorporate Moorish lines and Gothic elements. The best example is the Alhambra in Granada, Spain. As a revival style, it can be combined with additional styles such as Mission revival and Mediterranean revival.

40 Riverside Public Library, Special Collections Archives, Riverside YWCA Papers.

41 Riverside Art Museum, Archives; Laura L. Klure, *Let's Be Doers: A History of the YWCA of Riverside, California, 1906-1992*. (Riverside, CA: Riverside YWCA, 1992).

42 http://www.fccriverside.org/about/building.php.

43 Ibid.
44 Ibid.
45 "Santa Fe Depot and the Railroads," by Steven Shaw, City of San Bernardino California Website, http://www.sbcity.org/about/history/santa_fe_depot.asp (accessed February 1, 2016).
46 National Register of Historic Places Registration Form, San Bernardino, *Santa Fe Depot*.
47 Donald Duke. *Santa Fe . . . The Railroad Gateway to the American West, Volume 2* (San Marino, CA: Golden West Books, 1997); Lee Gustafson and Phil Serpico, *Santa Fe Coast Lines Depots: Los Angeles Division* (Palmdale, CA: Omni Publications, 1992).
48 National Register of Historic Places Registration Form, *Kelso Depot, Restaurant and Employees Hotel*, signed by National Park Service, June 20, 2001, and California State Historic Preservation Office, June 17, 2000; Duke, *Santa Fe . . .* Harvey Houses were comfortable, reliable hotel and restaurant accommodations for travelers along the Santa Fe railroad. Numerous Harvey Houses were built in the Mission Revival style, with adobe walls, arched colonnades, and tile roofs, from the Midwest to California.
49 National Register of Historic Places Registration Form, *Kelso Depot*.
50 "Kelso Depot: A Railroad Through the Desert Mojave Nation Preserve," United States National Park Service Website, http://www.nps.gov/moja/learn/historyculture/kelso-depot.htm (accessed January 10, 2016).

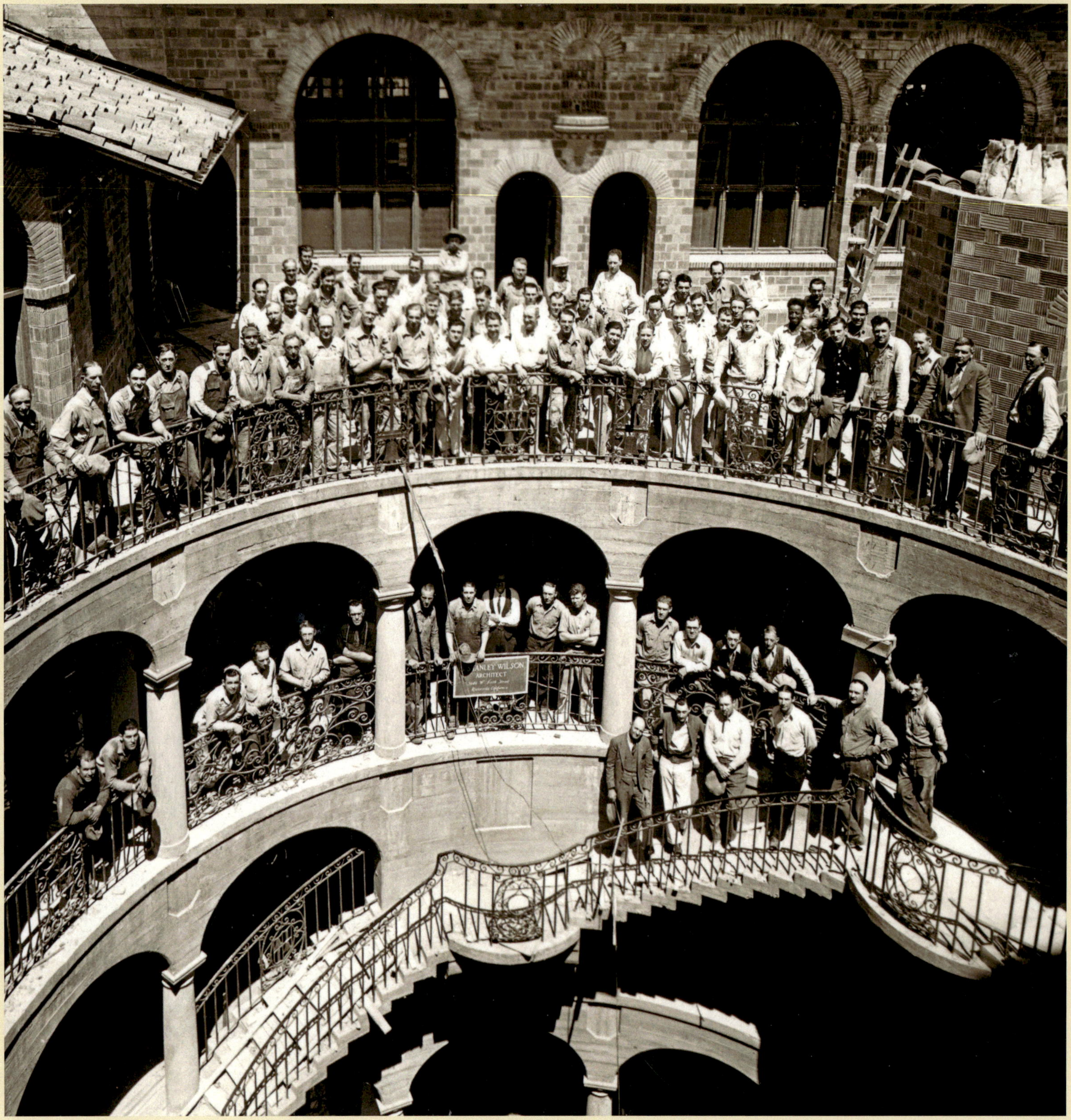
ANLEY WILSON
ARCHITECT

H. Vincent Moses and Catherine Whitmore

Building the Mission Inn

Frank Miller's Program to Create a Spanish Revival Oz in Riverside, California, 1903–1935

Both architecturally and in terms of boosterism, Frank Miller's Mission Inn . . . pushed Lummis's Spanish myth about as far as it could go, which in Southern California was very far indeed. . . . Beginning in the late 1890s, he embarked upon [thirty-five] years of architectural fantasizing, creating a Spanish Revival Oz: a neo-Franciscan fantasy of courts, patios, halls, archways, and domes, which he furnished with statuary, stained glass windows, and religious artifacts of Spain, Italy, and Mexico, gathered on pilgrimages abroad.

—Kevin Starr, *Inventing the Dream* (1985)[1]

FIGURE 1 Architect G. Stanley Wilson, standing second from right, top row, with construction crew Cresmer Manufacturing Co., ca. 1931, nearing completion of the International Rotunda, Mission Inn

Frank Augustus Miller's Spanish Revival Oz, the world-renowned Mission Inn National Historic Landmark, was not born Spanish. The hotel began in the late 1870s as an altogether Anglo-American hostelry. Captain C. C. Miller, Frank's father, opened the original hotel in an Anglo-style adobe, and a few modest American vernacular buildings as a winter resort for eastern and foreign visitors. He named it the Glenwood Cottages (fig. 2).

Frank Miller purchased the Glenwood in 1880, subsequently renaming it the Glenwood Tavern. By 1898, the ambitious young entrepreneur had set about to replace the original facilities with a modern multi-story hotel, initially of typical eastern hotel design. He failed to secure financing and scrapped the project. Meantime, however, Riverside had become the immensely prosperous home of Washington navel orange production, and the center of a rising Mission

Cult in Southern California vigorously promoted by Charles Fletcher Lummis. Navel orange prosperity and the Mission myth made Riverside an even more exotic and sought-after destination for well-heeled winter visitors. Miller never ceased efforts to fund a new hotel. Henry Huntington, owner of the Pacific Electric Railway, stepped to Miller's rescue in 1902, financed the new hostelry, and convinced Miller to commission Los Angeles architect Arthur Benton to design it. Benton rendered the new hotel in an audacious Mission Revival style, making a complete break with the original Anglo Glenwood Cottages. Miller's dream came to fruition May 8, 1903, when President Theodore Roosevelt personally presided over the grand opening of the New Glenwood, California's Mission Inn (fig. 3). He never looked back.

As Frank Miller proudly watched Riverside's Mediterranean Civic Center take shape in the late teens and 1920s, the visionary entrepreneur moved decisively and relentlessly to keep his hotel ahead of the pace. Through a constant construction program from 1911–32, involving multiple additions and renovations, Miller brought the original Mission Revival masterpiece into the complete and unsurpassed embodiment of both phases of the Spanish Colonial Revival (fig. 4). The transformation began in 1913, after Miller's grand tour of Europe, when he commissioned Myron Hunt to design the Spanish Wing, including the Spanish Patio, Spanish Dining Room, and Spanish Art Gallery. Built of reinforced slip-form concrete, the addition marked the dual advance of the Mediterranean Revival, and modern construction materials and methods at the Inn. Arthur Benton and local architect G. Stanley Wilson, who supervised construction of Hunt's Spanish Wing, designed the third and fourth floors. The north elevation features the Alhambra Court, Miller's ode to Granada. A row of spires, atop faux buttresses, infer minarets from a Spanish or North African mosque (fig. 5). The front façade of the Court is clad in hollow tile like the tile found at The Red Citadel itself (fig. 6).

Miller completed a final addition, the International Wing, between 1929 and 1932,

FIGURE 2 Glenwood Cottages, Riverside, ca. 1885, renamed The Glenwood Tavern by proprietor Frank Miller in the mid-1890s

FIGURE 3 Mission Revival arches, Seventh Street, Mission Inn, ca. 1915. Arthur B. Benton, architect

FIGURE 4 Cloister Wing, Mission Inn, ca. 1910. Arthur B. Benton, architect

with G. Stanley Wilson, Riverside's most prolific architect of the SCR, especially for schools and civic structures (fig. 1). Wilson was born in Bournemouth, England, and came to Riverside in 1895 with his family. Graduating high school in Riverside, he rose far and fast through the ranks of finish carpenter to self-taught architect via correspondence school courses and became a licensed architect in 1923. By the time Wilson got the commission for the Inn's International Wing, his firm had grown considerably. Wilson and the Inn benefited immeasurably from the arrival of Peter Weber, who brought with him immense artistic and architectural skills, and who had traveled to Europe and North Africa. Weber became Wilson's most important asset for designing in Mediterranean Revival style.

Weber's skills enabled Wilson to make his most significant contributions to the National Historic Landmark Mission Inn, especially the International Wing, which included the International Rotunda, the Court of the Orient, and the iconic Atrio of Saint Francis Addition. "The International Wing comprises the entire northwest corner of the Inn and was built between 1929–31 out of reinforced concrete. The walls, beams, floor slabs, and exterior walls

FIGURE 5 Author's Row, Mission Inn, 2016. G. Stanley Wilson, architect

FIGURE 6 North elevation, Mission Inn, ca. 1929, depicting proposed additions. G. Stanley Wilson, architect

are of reinforced concrete. The fourth-floor rooms are all of hollow tile construction with concrete roof framing supporting clay tile roofs," and "organized around three interior courtyards, all of which were ornately decorated in the spirit of international harmony." The six-story *Rotonda Internationale* (International Rotunda) constitutes the first court, rising the full height of the building, and open to the sky (fig. 8). The *Court of the Orient* makes up the second component of the wing, finishing with the *Atrio of St. Francis*, Wilson's opus in the Spanish Renaissance Revival Style, with its ornate Churrigueresque details (fig. 7). The whole building "varies in height from the four-story Mission Wings and seven-story Rotunda to the four commanding towers, Carillon, Carmel, Amistad and Agua, with many courts and elevations."[2]

The Atrio of Saint Francis drew rave reviews from the venerable *California Arts and Architecture*:

Architecturally, the Atrio of St. Francis will be considered the piece de resistance of Mission Inn. It might be the plaza of a small city of Mexico or Spain. The facade of the Chapel of St. Francis is the chief architectural feature. . . . Facing it from the entrance, one feels as if standing in front of a cathedral in a quiet plaza of Old Mexico. The Churrigueresque rich ornamentations, the rose windows, the coats of arms, the figures of saints in their niches—all are beautiful

FIGURE 7 Elaborate Churrigueresque entry surround with rose window, St. Francis Chapel and Atrio, Mission Inn, 2016. G. Stanley Wilson, architect

FIGURE 8 International Rotunda, Mission Inn, 2016. G. Stanley Wilson, architect

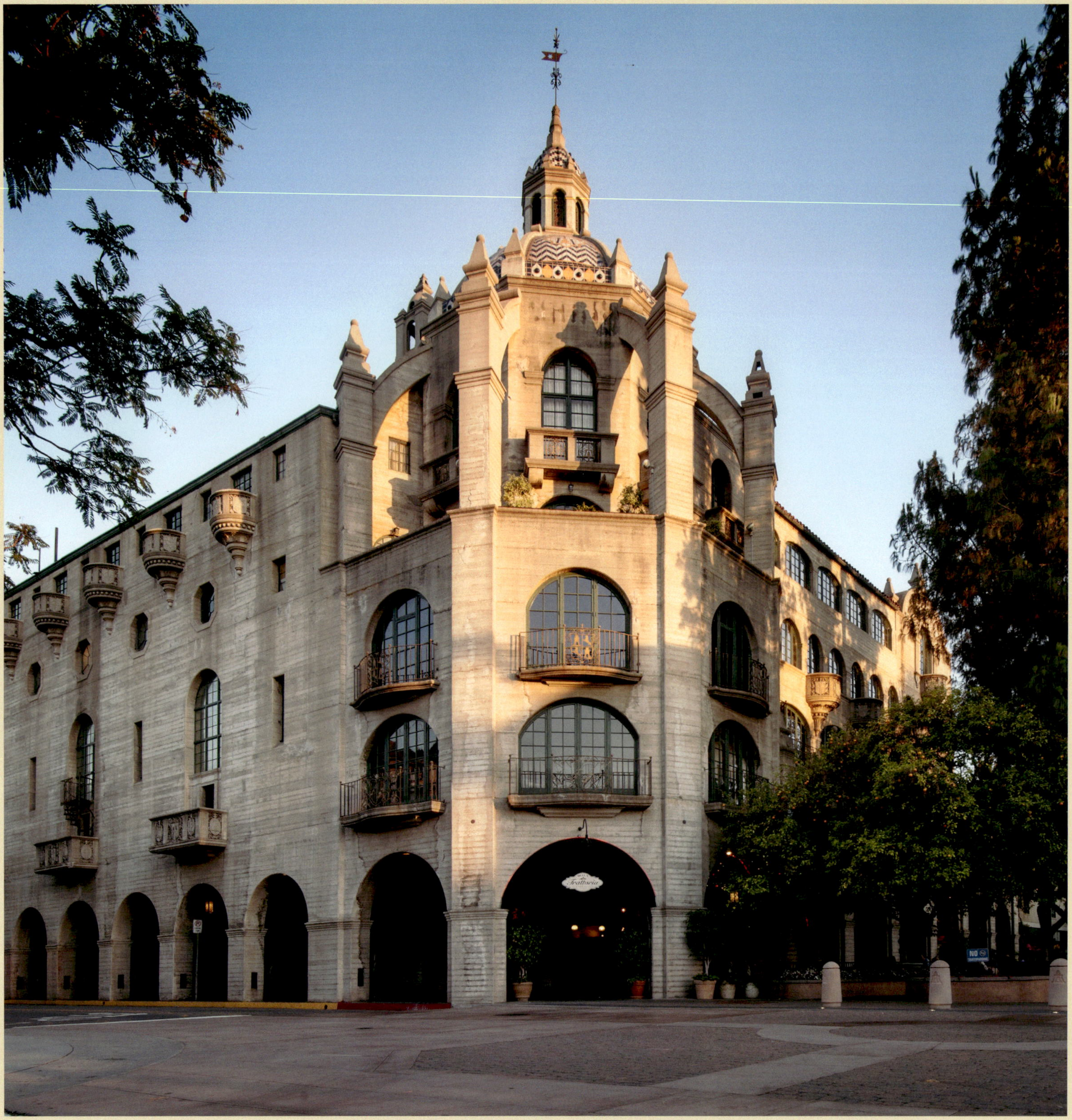

and all seem as if they must be of some bygone age. The proportions of facade and doorway and rose window are splendid. Huge sixteen-foot mahogany doors give entrance. The interior, dimly lighted, reveals its richness slowly to one entering from the brilliant sunlight of the Atrio. At the far end is the famous gold altar from Mexico. . . . Its surface and columns and figures have lost none of the lustre . . . they had two hundred years ago when the altar was made for the chapel of Marquis de Rayas at Guanajuato. Carved oak stalls of Renaissance design with medallions from an ancient monastery in Belgium occupy the sides of the chapel from the entrance to the chancel and above them, glowing and sparkling in all their color, are the Tiffany windows and mosaics, three on each side.[3]

The Atrio brought the Mission Inn to the pinnacle of the mature SCR, and to the forefront of the region's infatuation with the style (fig. 9).

At full build-out, the National Historic Landmark Mission Inn occupies a full city block at the heart of Riverside's downtown Civic Center, functioning once again as a five-star hotel, sponsoring a massively successful Festival of Lights every December that draws thousands of visitors to downtown Riverside.

Notes

1. Kevin Starr, *Inventing the Dream: California Through the Progressive Era*, (New York: Oxford University Press, 1985), 86.
2. City of Riverside Historic Property Inventory, 3649 Mission Inn Avenue, Architectural Description, Historic *Glenwood Mission Inn*, 2003 Survey, http://olmsted.riversideca.gov/historic/ppty_mtp.aspx?pky=5108. The Inn is designated Landmark #1 of the City, Contributor to the NR Seventh Street Historic District, and a National Historic Landmark of the United States.
3. Quoted in http://freepages.genealogy.rootsweb.ancestry.com/~npmelton/rvwils2.htm, George Stanley Wilson. Visit http://webmail.missioninnmuseum.org/collect_archi/archi_cont.htm, for a sterling short history and description of the architecture of the Mission Inn, including the Alhambra Court, ca. 1923-24.

FIGURE 9 Poured slip-form concrete and polychrome tiled Moorish dome, exterior, International Wing, Mission Inn at Sixth and Main Streets, 2016. G. Stanley Wilson, architect

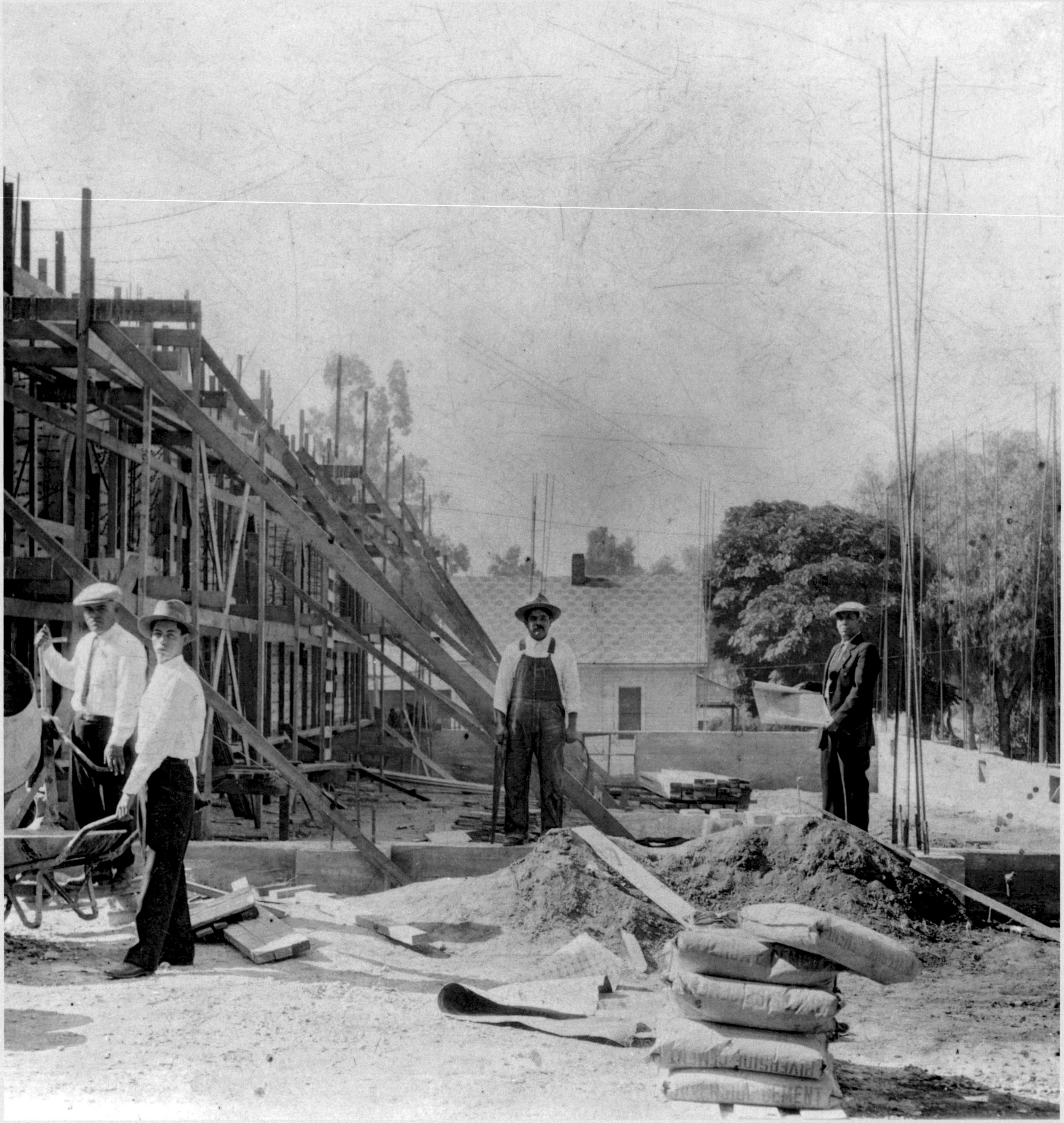

4

Carolyn Schutten

Voids of the Aggregate

Materializing Ethnic Mexicans in Mission Revival and Spanish Colonial Revival in the Inland Empire

December 25, 1916 ... Shovel crew consists of engineer, crane man, fireman, nightwatch and 7 Mexicans.

—Work log entry from California Portland Cement Company in Colton, California[1]

Rah! for the age of concrete!

—Redlands Citrograph, July 16, 1903

American Studies scholar Roberto Ramón Lint Sagarena, in *Aztlán and Arcadia: Religion, Ethnicity, and the Creation of Place*, notes that the ubiquitous presence of Mission and Spanish Revival architecture bespeaks a Southern California aesthetic that is both celebratory of and yet ultimately divorced from its Mexican history.[2] Spanish Colonial Revival architecture and the narratives of a romantic California past in the late nineteenth and early twentieth centuries assumed a continuity between Spain's colonial enterprise and American expansion, largely omitting ethnic Mexicans from the dominant SCR narrative.[3] Sagarena argues that there was "considerable synergy between historical narratives that naturalized American possession of the land and the creation of regional architecture." SCR architecture in Southern California flourished amid a romanticized past put forth by myriad Anglo boosters, builders, writers, architects, and real estate speculators, but it also emerged against the backdrop of the social and racial reordering post-Mexican-American War (1846–48) that cast Anglos as the new colonialists and redefined ethnic Mexicans as foreigners.[4]

FIGURE 1 Cement miners at California Portland Cement Company, Colton

In *California Vieja: Culture and Memory in a Modern American Place,* architectural historian Phoebe Kropp argues that colonial Spain gave way to American empire, and that regardless of delightful depictions of the Spanish era, Anglos "did not typically signal a willingness to embrace Mexicans or Indian Californians as fellow citizens."[5] Racial discourses, Kropp claims, were fundamental to the ways that Southern California Anglos defined local identity, boosted economic progress, built houses, interpreted history, and more.[6] Kropp notes that the leading local press acknowledged two groups in the region: Mexicans and Anglos, or "not Mexicans," masking sizable and significant differences among ethnic Mexicans.[7] "Ethnic Mexicans" in the early twentieth century might have included native-born *mestizos,* Spanish descendants, *genízaros* (Spanish indigenous captives), migrants from Central Mexico fleeing the Mexican Revolution, Mexican-Americans, and California Indians. Anglos marshaled "Spanish" to establish Europeanness, thus erasing Mexicanness in the local landscape.[8] Yet even as Anglo architects and developers lauded the beauty of "California Mission," "Spanish-American," or rarely "Mexican" architecture, Mission Revival and SCR architecture in the Inland Empire was in large part produced by ethnic Mexican common laborers working in building materials production and construction—and their contributions were particularly significant in clay and cement.

In the production of concrete—a key material in the construction of Mission Revival and SCR architecture—voids are the

FIGURE 2 Ethnic Mexican brick laborers and Native American local legend Willie Boy, subject of an infamous desert manhunt, standing second from left with wheelbarrow, Redlands

spaces in between the aggregate. Aggregates include sand, gravel, or crushed stone that, when mixed with water and Portland cement, comprise concrete (fig. 1). Numerous historical articles have been written on the effect of voids on the strength and durability of concrete.[9] Eliminating the voids does not necessarily produce the strongest concrete. Like the voids in concrete, the role of ethnic Mexicans in Mission Revival and SCR architecture has been largely omitted from the historical record, yet their labor and the production of building materials have played an integral part in the architectural history of the Inland Empire and Southern California. This essay intervenes in prevalent narratives of regional architecture and seeks to recover the history and contributions of ethnic Mexicans during the era that produced Mission Revival and SCR architecture (fig. 2).

In 1939, borderlands studies pioneer Herbert Eugene Bolton contradicted the Turnerian frontierist history of westward expansion, arguing that European-American expansion could be better understood as "the meeting place and fusing place of two streams of European civilization, one coming from the south, the other from the north."[10] However, historian Mario T. Garcia counters this statement, explaining that, though Bolton offered a wider lens with which to view Southwestern American history, he still possessed notions of "Spanish fantasy." The American Southwest after the United States conquest was not Spanish, but Mexican, Garcia argues. Alta California was "a mestizo society, one which had fused the Indian elements with the Spanish ones to create a distinct national culture."[11] High-ranking Californios and California natives were stripped of land, diminished in social standing, or relegated to working as common laborers in a relatively short period of time after America assumed control of California.[12] In the early twentieth century, Mexican nationals would flee from the Mexican Revolution, emigrate to California, and join other ethnic Mexicans working in agriculture or construction industries (figs. 3–4).

To understand the complexities of racial categories in Alta California before the Mexican-American War, one need only to look to the banks of the Santa Ana River to La

FIGURE 3   Construction of Burrage Mansion, Redlands

FIGURE 4 Construction of U.S. Post Office, Riverside

Placita de los Trujillos, also known as San Salvador, or to Agua Mansa in the areas just North of Riverside. In the 1840s, Lorenzo Trujillo led a group of *genízaros* from Abiquiu, New Mexico, and settled a 2,000-acre parcel near the Santa Ana donated by Don Juan Bandini. *Genízaros* were Native American captives of other tribes that were ransomed by the Spanish and traded as domestic slaves.[13] After Mexico's independence, *genízaros* were made Mexican citizens, though indigenous peoples were still at the very bottom of Mexico's very complex racial classification system. Still, Trujillo's *genízaro* settlements prospered in La Placita and Agua Mansa, and several European-American pioneers married *genízaro*, Spanish, or Californio women and had big families, including Cornelius Boy Jensen, Louis Robidoux, and Isaac Slover.[14]

The population of ethnic Mexicans in Southern California increased in the early twentieth century as families grew and as Mexican natives migrated north. In 1909, *The Mexican Mining Journal* commented on a recent bulletin by the Department of Commerce and Labor that native Mexicans that had been formerly working along the border of southwestern America were now in far flung regions throughout the United States, noting that 60,000 to 100,000 Mexican natives were entering the U.S. annually and "displacing Japanese, Greeks, and Italians."[15] Moreover, the violence and social displacement caused by the Mexican Revolution, along with the subsequent economic downturn in postwar Mexico, prompted hundreds of thousands of Mexican natives to travel north in search of solace and employment during the 1910s and 1920s. California's ethnic Mexican population, which had numbered only 51,137 in 1910 reached 368,013 by 1930.[16]

Despite Southern California's complex racial landscape, most ethnic Mexicans were categorized simply as "Mexicans" after the United States assumed control of the American Southwest—a term that would become derogatory and that was distinct from "Spanish."[17] Racial reordering in Southern California persisted into the early twentieth century, manifesting even in the building materials of Mission Revival and SCR architecture. Historian William Deverell, in *Whitewashed Adobe: The Rise of Los Angeles and the Remaking of Its Mexican Past*, argues that the use of adobe stood for the past, a "dark-skinned past."[18] Though it was the romance of the California missions that prompted the architectural turn to Mission Revival and later SCR, native adobe was maligned as an inferior building material. Charles Lummis in *Land of Sunshine* remarked in 1894 that the "old adobe hereabouts was not built so workmanlike as we prefer our homes."[19] Lummis argued that builders need not use "inevitable adobe," because it would require the importation of knowledgeable *adoberos* from Mexico. Lummis asks, "What, then, are the lessons the Superior Race [sic] might profitably learn from the adobe?"[20] Lummis recommended an "improved" adobe made of brick or cement, materials traditionally made by European-Americans. The Riverside Portland Cement Company (fig. 5) apparently agreed and admonished adobe in 1914 as the "most troublesome soil the contractor [had] to contend with."[21] Poignant examples of the inferiority of adobe were held up by Lummis and others, who pointed to the crumbling ruins of missions as proof positive of adobe's deterioration over time. Adobe meant backwardness to American arrivals, according to Deverell, "adobe was the unusable past."[22]

This aversion to the use of native building materials dovetailed the availability of local resources in Southern California. SCR, along with Mission Revival, was likely a strategic choice due to abundant mineral deposits in the Inland Empire and Southern California. In 1893, Governor H. H. Markham issued a report of the *Resources of California*, outlining the various industries already under way in California and extolling the virtues of the state's beauty and climate. North of Riverside in Colton, Slover Mountain, named after Isaac Slover of Agua Mansa, was already producing building materials. Markham reported that the marble columns of the California State Building at the Columbia Exposition had been quarried at Slover Mountain while also hinting at the possibility of developing other mineral resources throughout the Inland Empire.[23] In 1906, State Mineralogist Lewis Aubury published a detailed bulletin of mineral resources urging the development of local resources.

California offers exceptional opportunities to the investor, particularly in the many mineral substances of economic importance. A large percentage of these minerals is imported from other states or foreign countries, even though the raw materials are here, not only in abundance, but in many cases of superior quality, and only awaiting the capital necessary to develop them.[24]

The building industry in California was closely linked to the production of brick, tile, concrete, and cement stucco—all of which constructed and "materialized" Mission Revival and SCR structures.[25] Aubury reported that numerous mining operations were already under way in the Inland Empire, including the Taylor Brothers Brick Company in Redlands and Alberhill Coal and Clay in Riverside County near Elsinore.

Knowledge of brick production was imported from the East Coast and efficiently replaced native adobe in Southern California. Brick meant progress when compared to "dirty, ugly, unprogressive adobe," argues William Deverell. Brick stood for the "Anglo future" and was a metaphor for the racial turnover that occurred when California became America.[26] Transitions between building materials were

FIGURE 5 Riverside Portland Cement Company, Crestmore

"synonymous and simultaneous" with racial realities.[27]

In the Inland Empire, brick was used in Mission Revival architecture in the late nineteenth and early twentieth century much like adobe bricks with a cement plaster skin, and local clay production was key to pushing Mission revivalism forward.[28] Brick and roofing tile was produced for the Los Angeles Pressed Brick Company using clay deposits from Alberhill Coal and Clay in Riverside County. Bricks were shipped to other parts of Southern California, other states, and even Mexico as the demand for "Spanish style" accelerated. The Alberhill mine produced bricks for a Honolulu federal building and shipped bricks to the entire western United States and Mexico. Alberhill clay was also prized for pottery and decorative tile; Ernest Batchelder, a leader of the Arts and Crafts Movement, primarily used Alberhill clay for his decorative tiles.[29] Built in 1898 and designed by architect T. R. Griffith, the Mission Revival A. K. Smiley Library was constructed using ethnic Mexican labor and Redlands bricks by Taylor Brothers Brick Company. Built in Redlands in 1901, architect

FIGURE 6 Union Pacific Bridge under construction using California Portland Cement Company Colton Plant, Riverside

Charles Brigham's Mission Revival Burrage Mansion also used Taylor bricks and local Mexican workers (fig. 3). C. P. Hancock & Son (fig. 9) produced brick with the help of ethnic Mexican laborers and constructed the Mission Inn Annex, the YMCA, the Robert Lutz residence on the corner of Mission Inn and Redwood Avenues, a Spanish Mediterranean Revival home on Larchwood Avenue, and the St. Francis De Sales Church in Riverside.[30] Local clay was also used to make roofing tile and decorative tile for Mission Revival and SCR structures throughout the Inland Empire. Sunset Tile Company in Redlands, for example, supplied the clay tile and roofing for Riverside's YWCA, now the Riverside Art Museum, designed by architect Julia Morgan.

Though clay roofing tile and decorative tile was crucial to the Mission Revival and SCR aesthetic, the brickmaking industry faltered by the 1920s, as concrete began to overtake brick as the preferred building material in the Inland Empire. Masonry block was already supplanting brick usage, and the myriad uses and low cost of cement made brick's decline inevitable. Moreover, the area's rich mineral deposits led to a thriving construction industry in Southern California and eliminated a need for imported materials. Imported Portland cement from Europe dropped nearly 3 million barrels in 1891 to 307,00 by 1910, while domestic Portland cement rose from 454,000 in 1891 to a staggering 76.5 million barrels in 1910. In 1920, the California State Mining Bureau reported that, due to the increased demand for building materials in 1919, cement plants were running at capacity.[31] In Los Angeles in 1918, 4.6 million dollars were spent on cement structures, and by 1923 that number had soared to nearly twelve million dollars.[32]

Advocates of concrete were quick to note the plastic qualities of cement and its suitability for Mission Revival and SCR architecture. Architect William Price stated in a paper read before the Association of American Portland Cement Manufacturers in 1909:

> But in a material so plastic the forms of openings and mouldings may be expected to vary much from those necessary to an architecture dependent on arches and lintels. There is more to be learned in the Spanish and the Mexican varieties of architecture, than any other accepted types. Their plastered walls, tile roofs and wall copings suggest concrete more than they do brick, and their domes and curved pediments are already suggestive of plastic rather than block construction.[33]

Concrete was lauded widely for its strength and its ability to withstand fire and floods, but Price impressed upon cement producers that

FIGURE 7 Construction of Our Lady of Guadalupe Shrine, Eastside, Riverside

FIGURE 8 Adobe laborers during the reconstruction of the San Bernardino Asistencia, Redlands

"Spanish and the Mexican varieties of Spanish" architecture were best suited to the plastic uses of concrete.[34] Cement proponents argued that its durability and plasticity not only outpaced buildings of adobe but also of brick and stone.[35] Price asserted that using concrete as merely a foundation for brick or stone was a "sham" and advocated a move to curving and molded cement that spotlighted the medium.[36] This sweeping move to cement as a dominant building material obviated the need for cement stucco "whitewashing" of brick, and the Inland Empire saw the construction of numerous SCR buildings with exteriors of reinforced concrete.[37] Ample limestone resources and the considerable plastic qualities of concrete may have contributed to the decline in Mission Revival architecture and the expansion of SCR as concrete found a solid audience.

Editor of *Pacific Coast Architect* Harris Allen remarked that by 1926 architects had been "coming closer and closer to the achievement of that Spanish atmosphere which was the glory of early California." Urban studies scholar Albert Fu observes that Allen emphasized architectural achievement, not rediscovery. Interpreting the "Spanish idea," according to Fu, meant besting early methods and inventing increasingly superior architecture. During the construction of Riverside's First Congregational Church in 1912, *Southwest Contractor* noted that the "elaborate and complicated" designs for the church by architects Myron Hunt and Elmer Grey and architectural drafter Henry L. A. Jekel were "largely adapted from a number of old [sixteenth and seventeenth century] Mexican churches."[38] The concrete subcontractor Fred Schupach attracted considerable local attention for "the best type of artificial stone work that [had] ever been done" in the Inland Empire, underscoring both the possibilities of cement as well as Anglo skill in reproducing a superior Mexican-style architecture in the region.[39]

Local cement producers employed full-time chemists, claiming that "[c]oncrete does not belong to the class of work that can be done by unskilled labor" and emphasizing the need for Anglo science and expertise.[40] However, cement already had a long history in the Inland Empire. After the Bandini Donation, the ethnic Mexican and *genízaro* settlers at Agua Mansa were already burning limestone to make plaster and cement for adobe structures, and by 1860, some primitive lime kilns were established at Slover Mountain.[41] In 1891, Slover Mountain, with its "limitless" supply of high quality limestone, was later said to be discovered by Anglo "farsighted businessmen," and California Portland Cement Company in Colton was incorporated later that year; it was the first cement company west of the Mississippi River.[42] The Colton plant supplied cement for the Union Pacific Bridge in Riverside

FIGURE 9 C. P. Hancock and Son brick workers, Riverside

(fig. 6), the Los Angeles Coliseum, Hoover Dam, and a Churrigueresque band shell for Fleming Park in Colton designed by Lloyd Wright.[43]

Riverside Portland Cement Company announced its debut at Crestmore near Riverside in December 1909 and would, along with the Colton cement plant, go on to play a critical role in the development of the Inland Empire SCR architecture.[44] In an article extolling the virtues of Mission style architecture, an anonymous reporter stated in 1909 that it would be "a dreadful pity to consider anything but a mission building" and that there was "nothing in classical architecture for Riverside but unfavorable comparison, while there is nation-wide fame in a consistent grouping of buildings in the historical, romantic style of native California architecture."[45] The article was adjacent a conveniently placed article announcing that the Crestmore plant would carry the Riverside name, noting that every barrel and bag of cement would be emblazoned with "Riverside" and would, like citrus crate labels, advertise the city as well as the cement.[46] Boosters made clear ties between regional architecture and local cement production. The Crestmore plant produced cement for the construction of the Riverside Municipal Auditorium, the California Baptist University core buildings, and numerous other SCR structures; local architects such as G. Stanley Wilson, Henry L. A. Jekel, and others favored structural concrete methods using Riverside cement.[47]

With the cement industry poised to take off, there was an immediate need for labor, and ethnic Mexicans were often preferred. An article on Mexican labor in the United States in 1909 stated that, "Mexicans are somewhat indolent and irregular, not ambitious, not particularly strong, but are fairly intelligent, orderly, and most of all, work for low wages, so that they compete in occupations paying the

FIGURE 10 Construction of St. Anthony's Church for the Casa Blanca neighborhood, Riverside

lowest wages and do not, therefore, antagonize native [European-American] workmen."[48] Ethnic Mexicans routinely made less money than Anglo workers. In 1928, construction supervisor W. H. McNeeley reported that all of the Anglo workers at California Portland Cement Company in Colton made at least double the 30 cents per hour that "Mexican common laborers" earned.[49] Only one Mexican concrete mixer, Preciliano Vasquez, made more at 35 cents per hour.[50]

The working conditions for cement miners were perilous. Cement dust reportedly caused bronchitis, skin diseases, influenza, diseases of the eyes, nose, throat, and ears, rheumatism, and tuberculosis.[51] Dr. George E. Tucker examined 956 cement workers—282 of whom were "Mexican"—from Riverside Portland Cement Company in October 1913. Tucker found that one in thirty-three of those laborers had deformities of the hands, arms, fingers, toes, chest, and shoulders, and several were missing fingers or toes, and one in forty-five workers were partially deaf due to cement mining and production.[52] Numerous workers had swollen lymph nodes or tonsils related to dust. Ethnic Mexican cement workers, Tucker noted, were also more likely to be single than American laborers, though Anglos were nearly twice as likely to have gonorrhea.

Cement mining and production was hazardous and sometimes fatal. Beyond just lost or maimed fingers and toes, some men had legs and arms blown off and some died of shock; others were buried alive, fell, suffered from neurasthenia, or shell shock, due to dynamite explosions.[53] The Riverside press reported on April 24, 1913, that eleven miners had lost their lives in a premature blast of five hundred pound charges of dynamite at the Riverside Portland Cement Company at Crestmore. Several of those miners were reportedly blown high into the air while the others perished underground. Nine of those workers were ethnic Mexican: six Mexican drillers, one Spanish driller, one Yaqui Indian driller, and one Mexican laborer.[54] It was reportedly several months before the Crestmore plant could find all of the bodies. Injuries sustained in cement production might also affect employment of laborers. In January 1927, the superintendent of the California Portland Cement Company in Colton reported that employees Julio Rivera, Francisco Galvan, Guillermo Flores, and Jesus Gonzalez were laid off "on account of only having the use of one eye."[55]

Native Americans also worked in masonry and construction on significant Mission Revival structures in Riverside. In the late 1930s, five major Mission Revival buildings at Sherman Institute in Riverside were erected entirely by Native American labor in the biggest construction activity in the history of the school.[56] The federal government established Sherman Institute on Magnolia Avenue in Riverside in 1902 as an off-reservation industrial boarding school meant to transform American Indian students into "productive" laborers.[57] The aim of the school, according to founder Richard Pratt, was to "kill the Indian and save the man" and facilitate assimilation.[58] Students were removed from their reservation homes and taught English and common labor work, such as agriculture, construction, and masonry. The Mission Revival architecture of the campus was designed to "give the impression of thick adobe bricks" and placed the students in a context that referenced California's past as well as the enormously popular Helen Hunt Jackson's novel, *Ramona*. Two of Sherman Institute's buildings were even named "Ramona Home" and "Alessandro Lodge."[59] Sherman Institute was closely linked to the Anglo development of Riverside,

FIGURE 11 Casa Blanca School designed by G. Stanley Wilson for the Casa Blanca neighborhood, Riverside

and boosters who sought to capitalize on the area's romanticized Spanish past. Frank Miller, founder of the Mission Inn, was instrumental in bringing the school to Southern California and also championed the construction of the streetcar route for the Mission Inn that included a stop at the Sherman Institute in order to create an added "attraction" for tourists who wanted to see Indians at work in a mission-like setting.[60]

Ethnic Mexican and Native American laborers were also used as a kind of tourist attraction during the reconstruction of the San Bernardino Asistencia in Redlands (fig. 8). Archivist

FIGURE 12 Our Lady of Guadalupe Shrine, Eastside, Riverside

and historian Nathan Gonzales argues that George Beattie ignited local preservation efforts in order to establish a tangible link between the Redlands area and boosters' romanticized Spanish past.[61]

The former Spanish *estancia* was established in 1819 as an outpost for cattle grazing for the Mission San Gabriel, but by the 1920s, the adobe structure was in ruins. The newly formed San Bernardino County Historical Society rallied restoration efforts in 1926, and construction was completed after the Depression in 1937 with funding from State Emergency Relief Administration and Works Progress Administration. In order to ensure that the project was "authentic," local California Indians and ethnic Mexican laborers were hired to build the structure using traditional methods of adobe and plaster.[62] These "skilled workman of Indian blood" painstakingly created adobe bricks and hefted each one up a ladder precariously balanced on the head in an effort to promote roadside tourism in Redlands during construction.[63]

The living conditions of ethnic Mexican laborers working in clay and cement in the early twentieth century were not unlike the Mission era in Southern California, according to William Deverell.[64] Brickmakers were housed adjacent mining and production operations at the Simons Brick Company in Los Angeles and also at Alberhill Coal and Clay near Elsinore in the Inland Empire. Critical components to brickmaking included ample quality clay deposits, a convenient railroad connection, cheap land, and a labor force of unskilled, cheap Mexican workers, Deverell states.[65] Brickmaking was labor intensive and comprised hundreds of ethnic Mexican workers; whole "all Mexican" communities sprouted up around the clay mines. Deverell likens these ethnic Mexican settlements to the California mission system, arguing that the Franciscan Indians had been replaced by Mexican industrial laborers, "now kept in place by the contingencies of a restricted labor market and attractions offered by paternalistic employers."[66]

Rather like the "paternalistic" structure of brick mining and manufacturing in Southern California, concrete laborers lived in a boarding house just near the cement plants at Crestmore and Colton in the early twentieth century. Though not much is known about the boarding house at the Colton plant, the Riverside Portland Cement Company worker's barracks were designed by architect Irving Gill. Gill was associate architect to Bertram Goodhue for the first building in Balboa Park for the 1915 Panama-California Exposition in San Diego, and he was a seminal figure in early modern architecture and a visionary in concrete who notably developed the "tilt-slab" construction method. Gill's rectangular design for some 200 ethnic Mexican laborers included an abundance of trees and a central garden and resembled the linear quarters for soldiers or priests in California Mission architecture. Unlike the boarding house for single men in Colton, the Crestmore barracks were designed for families and furnished with individual rooms with stoves and running water, a central pavilion for music, flower and food gardens, eucalyptus trees, and a first aid facility.[67] The Riverside Portland Cement Company was held in high regard by the Department of Labor and Commerce; the compound even offered English lessons on site, which were taught by teenager Flora Robidoux, great-granddaughter to Louis Robidoux.[68] The barracks were primarily for ethnic Mexicans as cement laborers of "other nationalities" reportedly lived in town.[69] Crestmore even had its own romantic tragedy that was likened to Helen Hunt Jackson's *Ramona*.

Lucy Villanuevo, a fourteen-year-old beauty known as the "Belle of Crestmore," was abducted in 1913. In a frenzy, her father sought out her beau, Atilano Cano, and killed him, only to discover that he had murdered the wrong man. Villanuevo's wife revealed that Lucy had in fact eloped with Jesus Mendosa while Villanuevo was still at large.[70] Villanuevo was captured three years later and sentenced to life imprisonment.[71]

Cement mining workhouses were usually run by a foreman, who policed the house and enforced strict rules which included a curfew and teetotaling. The relationship between foremen and Mexican workers was often volatile and more than once led to mayhem or murder. In 1911, José Martinez of California Portland Cement Company in Colton stabbed foreman James Canterbury; the altercation reportedly started because Martinez "verbally resented" an order given by the foreman.[72] At the Riverside Portland Cement Company, Frank Salinas, a "Mexican" according to press, returned after being fired by a foreman with an automatic weapon. Salinas was tackled and handcuffed and taken to jail.[73] Foreman S. C. Harrison shot and killed cement worker Tomas Negrete at Crestmore for breaking curfew and drinking. Forty-five minutes after lights out, Harrison found Negrete and some other workers drinking wine and singing to guitar music.[74] When the foreman started to pour out the wine, Negrete "attacked" and was killed in self-defense.[75]

Conflicts between ethnic Mexican cement workers and foremen often led to strikes.[76] In August 1917, after a physical altercation with foreman James Ferrell, one hundred workers launched a strike at the California Portland Cement Company in Colton—allegedly with support from the Industrial Workers of the World (IWW).[77] Local cement workers organized the *Trabajadores Unidos* union and mobilized a two-month strike against the Colton cement company. The strike was in protest of a cut in ethnic Mexican wages by the Colton plant, "whose rationale for the wage cut was that Mexicans were making too much money."[78] *Trabajadores Unidos* eventually won wage concessions, union recognition, and reinstatement of workers.[79] In July 1918, one hundred twenty-five "Mexican roustabout laborers" launched a strike in order to earn the right to choose their own foreman.[80]

Despite these tensions, the cement industry over time drew whole families and multiple generations into the business, affording ethnic Mexicans opportunities to establish families, unions, businesses, and communities. Gene Juarez, a former employee of Colton's California Portland Cement Company in Colton for forty-three years, boasted four generations of laborers. His grandfather, Miguel Carreon, arrived from Mexico in 1923 and began work the next day at the Colton plant. His great uncle, Martiniano Carreon, put in forty years of service, and his father, Jesus Juarez, worked there for forty-two years after World War II. Retired cement worker Gary Thornberry married Esther Alvarez, whose father worked at the Colton plant and whose family descended from the *genízaro* settlement in Agua Mansa.[81] Her grandfather, Estanislacio Pietro, worked at the California Portland Cement Company in Oro Grande and managed a cantina and cottages that catered to cement workers. Other ethnic Mexican families with multiple generations and decades of service include the Guerreros, the Campas, and the Razos.[82] *Trabajadores Unidos*, the union organization forged in South Colton, played an important role in workers' lives, according to Juarez.

Colton cement plant employees began to populate a *barrio* in South Colton in the early twentieth century. Over time, ethnic Mexicans purchased tiny lots and built small wood frame homes with available materials, such as discarded lumber and corrugated metal.[83] Juarez recalls that homes in South Colton were constructed with old wood siding from rail cars and the streets were unpaved until the 1960s.[84] The structures were not connected to a central sewer system, and his grandfather's home had an outhouse for decades. By 1913, though, South Colton had an underground Spanish-language academy, and community organizations, included a mutual aid society, a committee for celebrating Mexican national holidays, and a women's Blue Cross society.[85] A number of small businesses, including bakeries, groceries, restaurants, pool halls, taverns, clothing, and butcher shops sprang up in South Colton, creating a thriving ethnic Mexican community.

Ethnic Mexican *barrios* in the Inland Empire during the early twentieth century were mostly devoid of Mission Revival and SCR materials and details. The vast majority of ethnic Mexican homes in South Colton, and Casa Blanca and Eastside in Riverside, were small wood frame houses on dirt streets that were lacking in sewage. A scant number were reportedly constructed with "adobe," but none contained the materials or the architectural flourishes of Mission Revival or SCR architecture.[86] Any Mission Revival or SCR structures meant for *barrios* were usually constructed entirely by ethnic Mexicans, such as the St. Anthony Church in Casa Blanca, Riverside (fig. 10), and the Our Lady of Guadalupe Shrine in Eastside, Riverside (fig. 12).[87] Ethnic Mexicans also built the Community Settlement House in Riverside, a Progressive Era philanthropic organization that taught Mexican immigrants English and trained them for domestic labor jobs.

However, when Mexican *barrios* needed public buildings, those contracts were usually given to Anglo architects but perhaps without the same measure of ornamentation. Celebrated Inland Empire architect G. Stanley Wilson, designed several notable examples of SCR architecture, including large portions of the Mission Inn, the Aurea Vista Hotel, and the Casa De Anza Motel. Wilson's design of the Casa Blanca School in Riverside (fig. 11), however, hints at SCR but is spare in architectural flourishes. Historian David Torres-Rouff notes that the façade, "with its angularity, shed roofed entrance, and exposed concrete material, speaks the modernist language of functionalism and cost-efficiency."[88]

The legacy of these largely invisible ethnic Mexicans in the construction of the Inland Empire landscape is difficult to trace, but several attempts have been made at materializing those laborers through oral history and historic preservation projects. Native Riversider Mary Jimenez Sosa Pasillas recalls that her grandfather Zenon Sosa worked for Riverside Portland Cement Company and also a brick company in downtown Riverside.[89] Sosa Pasillas remembers seeing her grandfather on the construction site for Our Lady of Guadalupe Shrine in Riverside (fig. 7).[90] Lupe Avila's father came from Mexico in 1900 and helped build numerous structures in Riverside, including the Community Settlement House; he also worked in adobe. Monica Arce's great-great grandparents were a part of the Trujillo expedition in 1843 and lived in Agua Mansa. Her grandfather was *genízaro* and was sent to Sherman Indian School after the family was "forced off the land when the government built the Prado Dam."[91] The Arce family lived in ethnic Mexican *barrios* afterward and were segregated and discriminated against, though they can trace their roots to before 1776 in San Juan

Capistrano and were listed in the 1928 Bureau of Indian Affairs census.[92] Monica's great grandfather Dan Arce was born across the Santa Ana River in San Salvador near Riverside and worked at Alberhill Coal and Clay.[93]

Deverell asserts that in Southern California the "color of brickwork [was] brown," and indeed numerous laborers that produced brick and concrete in the Inland Empire were ethnic Mexicans. The vast majority of Mission Revival and SCR architecture in the Inland Empire was supplied materials by Riverside Portland Cement Company and California Portland Cement Company in Colton, and numerous structures were constructed with clay from Alberhill Coal and Clay in Elsinore, Taylor Brothers Brick Company, and C. P. Hancock & Son. And very likely all Inland Empire Mission Revival and SCR structures were built by countless ethnic Mexicans—Mexican-Americans, *genízaros*, Native Americans, Mexican immigrants, and native Mexican Californians.

In 1994, the city of Riverside's Planning Department approved the development of seventeen homes in the Casa Blanca Redevelopment Area along Evans and Pliny Streets in Riverside.[94] Built in 1995 by Corydon Construction, these bungalows, with their red roof tiles, fireplaces, and spacious patios, brought SCR architecture to the 100-year-old ethnic Mexican neighborhood. In a *barrio* of wood frame houses, these modest SCR bungalows appear strikingly authentic and beckon the inclusion of Casa Blanca into Riverside's historical narrative.

By exploring the role of ethnic Mexicans in the social and material history of Mission Revival and SCR architecture in the Inland Empire, the story of concrete materializes. Cement would later in the twentieth century dominate architecture, peaking during the era of urban renewal in the 1960s when modernizing often meant bulldozing brick and other historic buildings. Cement would play a key role in shrouding structures in monolithic white and the construction of roads and highways that often separated ethnic Mexican and Anglo neighborhoods. However, Southern Californians are still intoxicated by regional history and identity and continue to chip away at the concrete. Citrus Belt Savings in Riverside attached a cement curtain wall over its façade in 1961. Some fifty years later, construction workers discovered a stunning Churrigueresque- style structure on Market Street that was designed by architect Stiles O. Clements in 1926 for the Riverside Finance Company. That structure today fittingly houses the Riverside Community College District's Center for Social Justice and Civil Liberties.

Notes

1. Work Log Entry California Portland Cement Company, December 25, 1916, Call Number: 126, Box 1, Folder 1, Ian A. Smith Papers Collection, Special Collections & Archives, University of California Riverside, Riverside, California.
2. Roberto Ramón Lint Sagarena, *Aztlán and Arcadia: Religion, Ethnicity, and the Creation of Place* (New York: New York University Press, 2014), 1–2.
3. Ibid., 2.
4. Phoebe S. Kropp, *California Vieja: Culture and Memory in a Modern American Place* (Los Angeles, University of California Press, 2006), 8.
5. Ibid., 4–5.
6. Ibid., 7.
7. Ibid., 9–10.
8. Ibid., 10.
9. For example, see Duff Andrew Abrams, *Design of Concrete Mixtures: Vol. 1* (Chicago: Structural Materials Research Laboratory, Lewis Institute, 1919).
10. Herbert Eugene Bolton, *Wider Horizons of American History* (Notre Dame: University of Notre Dame Press, [1939] 1967), 98. For more on Turner's frontierism, see Frederick Jackson Turner, *The Significance of the Frontier in American History* (London: Penguin, [1893] 2008).
11. Mario T. Garcia, "A Chicano Perspective on San Diego History," *The Journal of San Diego History* 18, no. 4 (Fall 1972), http://sandiegohistory.org/journal/1972/october/chicano-2/.
12. Ibid.
13. See also Russell M. Magnaghi, "Plains Indians in New Mexico: The Genizaro Experience," *Great Plains Quarterly* (1990): 86–95.
14. Susan Straight, "Agua Mansa: Californio Roots in the Inland Empire," *KCET.org* (September 21, 2011).
15. "Mexican Labor in the United States," *The Mexican Mining Journal* VIII, no. 4 (April 1909): 23.
16. David Torres-Rouff, "Becoming Mexican: Segregated Schools and Social Scientists in Southern California, 1913–1946," *Southern California Quarterly* 94, no. 1 (2012): 99.
17. Kropp, *California Vieja*, 9.
18. This research was guided by William Deverell, *Whitewashed Adobe: The Rise of Los Angeles and the Remaking of Its Mexican Past* (Los Angeles: University of California Press, 2004), 133.
19. Charles Lummis, "The Lesson of Adobe," *Land of Sunshine* (December 1894): 65.
20. Ibid.
21. Riverside Portland Cement Company, *Good Concrete: A Manual for the Rational Use of Portland Cement* (Los Angeles: DeWit and Co., 1914), 75.

22 Deverell, *Whitewashed Adobe*, 134.
23 H. H. Markham, *Resources of California* (Sacramento: Superintendent State Printing), 58.
24 Lewis E. Aubury, *The Structural and Industrial Materials of California: Bulletin No. 38* (Sacramento: Superintendent State Printing, 1906), 5.
25 This essay builds on sociologist Albert Fu's notion of "materializing" SCR as a research methodology for excavating the social and material history of ethnic Mexicans in the region's production of SCR. Albert S. Fu, "Materializing Spanish-Colonial Revival Architecture: History and Cultural Production in Southern California," *Home Cultures* 9, no. 2 (2012): 159.
26 Deverell, *Whitewashed Adobe*, 135.
27 Ibid.
28 Ibid., 139.
29 Robert Winter, "Ernest Batchelder (1876–1957)," *California Tile—The Golden Era 1910–1940* (Atglen, PA: Schiffer Publishing Ltd., 2003).
30 "Last Summons for Claude P. Hancock," *Riverside Daily Press*, March 2, 1931, 7.
31 W. Burling Tucker, "Los Angeles Field Division," *Report XVII of the State Mineralogist: Mining in California in 1920* (San Francisco: California State Mining Bureau, January 1921), 270
32 Fu, "Materializing Spanish-Colonial Revival," 169.
33 William A. Radford, *Cement Houses and How to Build Them* (Chicago: The Radford Architectural Company, 1909), 12.
34 Ibid.
35 Ibid.
36 George E. Thomas, *William L. Price: Arts and Crafts to Modern Design* (Princeton: Princeton Architectural Press, 2000), 131–33.
37 See: Deverell for an extensive discussion of "whitewashing" and the role of ethnic Mexicans in brick production in Los Angeles.
38 *Southwest Contractor*: 9, no. 24 (October 19, 1912): 9.
39 Ibid.
40 Ibid., 8.
41 Manuscript by Ian A. Smith, undated, Call Number: 126, Box 3, Folder 9, Ian A. Smith Papers Collection, Special Collections & Archives, University of California Riverside, Riverside, CA.
42 Ibid.
43 Chuck Wilson, *Quality Unsurpassed 1891–1991* (Glendora: California Portland Cement Company, 1991), 82–85.
44 "Riverside Brand Cement on Market December 15," *Riverside Daily Press*, November 23, 1909, 10. Both plants changed names over the years. To avoid confusion, they will be referred to by their respective locations in Crestmore and Colton.
45 "Mission or Classic," *Riverside Independent Enterprise*, October 24, 1909, 4.
46 "Riverside Brand," ibid., 4.
47 Interview with H. Vincent Moses, September 15, 2015.
48 "Mexican Labor in the United States," 23.
49 Letter from W. H. McNeeley to Warren Trevell, California Portland Cement Company, December 3, 1928, Ian A. Smith Papers Collection, Call Number: 126, Box 2, Folder 10, Special Collections & Archives, University of California Riverside, Riverside, CA.
50 Ibid.
51 Albert E. Russell, M.D., "The Effect of Cement Dust Upon Workers," *The American Journal of the Medical Sciences* 185, no. 3 (1933): 338.

52 George E. Tucker, "Physical Examination of Employees Engaged in the Manufacture of Portland Cement," Conference Paper for the Public Health Association, November 30–December 4, 1914, 568, https://www.ncbi.nlm.nih.gov/pmc/articles/PMC1286625/pdf/amjphealth00106-0068.pdf.
53 Ibid.
54 "Eleven Men Killed in Dynamite Blast," *Riverside Enterprise*, April 24, 1913, 2.
55 Letter from Superintendent to W. H. McNeeley, California Portland Cement Company, January 14, 1927, CPCC Documents, Box 3, Mexican Labor Letter, Chemist, CalPortland Historical Archives Collection, Glendora, California.
56 "Five New Buildings Change Sherman Institute Outlook," *Riverside Daily Press*, January 1, 1938, 20.
57 For more on the Sherman Institute, see Clifford E. Trafzer, Matthew Sakiestewa Gilbert, and Lorene Sisquoc, eds., *The Indian School on Magnolia Avenue: Voices and Images from Sherman Institute*, (Corvallis: Oregon State University Press, 2012).
58 Nathan Gonzales, "Riverside, Tourism, and the Indian: Frank A. Miller and the Creation of Sherman Institute," *Southern California Quarterly* 84, no. 3/4 (2002): 194. For more on Indian labor at Sherman Institute, see Kevin Whalen, *Native Students at Work: American Indian Labor and Sherman Institute's Outing Program, 1900–1945* (Seattle: University of Washington Press, 2016).
59 Gonzales, "Riverside, Tourism," 214.
60 Ibid., 194.
61 Nathan Daniel Gonzales, "Visit Yesterday, Today: Ethno-tourism and Southern California, 1884–1955" (PhD diss., University of California Riverside, 2006), 129.
62 Ibid., 141.
63 Ibid.
64 Deverell, *Whitewashed Adobe*, 142.
65 Ibid., 137.
66 Ibid., 142.
67 How Cement Plants Cares for Its Mexican Laborers," *Riverside Daily Press*, April 10, 1914, 3.
68 "Girl in Her Teens Teaches Old Men," *Riverside Daily Press*, November 25, 1913, 8. Robidoux commented, that the cement workers were "men of giant strength who [were] taking up the work of children."
69 "How Cement Plant Cares for Its Mexican Laborers," 3.
70 "Abducted Killer and Girl Still Missing," *Riverside Enterprise*, October 12, 1913, 8.
71 "Villanuevo Found Guilty of Murder," *Riverside Daily Press*, June 13, 1919, 3.
72 "Stabbing Affray May Mean Murder," *Riverside Morning Enterprise*, November 25, 1911, 3.
73 "Mexican Ran Amuck With Automatic Weapon," *Riverside Daily Press,* March 29, 1913, 1.
74 "Night Watchman Kills in Self-Defense," *Riverside Enterprise,* July 18, 1916, 3.
75 Ibid.
76 For more on early ethnic Mexican labor, see Juan Gómez-Quiñones, "The First Steps: Chicano Labor Conflict and Organizing 1900–1920," *Aztlán*, no. 1 (1973).
77 "I.W.W. Influence at Cement Plant," *Riverside Daily Press,* August, 9, 1917, 7.
78 José Pitti, Antonia Castaneda, and Carlos Cortes, "A History of Mexican Americans in California," *Five Views: An Ethnic History Site Survey for California* (Sacramento: Office of Historic Preservation, 1988), 242.
79 Héctor L. Delgado, "Unions and the Unionization of Latinas/os in the United States," in *Latinas/os in the United States: Changing the Face of América* (New York: Springer, 2008), 372.

80 "Cement Strike On," *Riverside Daily Press,* July 13, 1918, 3. The California State Mining Bureau reported cryptically in 1920 that mining labor was in short supply and that the labor that "was to be secured was inefficient." W. Burling Tucker, "Los Angeles Field Division," *Report XVII of the State Mineralogist: Mining in California in 1920* (San Francisco: California State Mining Bureau, January 1921), 270.

81 Interview with Gary Thornberry, December 5, 2016.

82 Interview with Gene Juarez, December 12, 2016.

83 José Pitti, Antonia Castaneda, and Carlos Cortes, "Historic Site: South Colton," *Five Views*, 242.

84 Ibid.

85 Ibid.

86 City of Riverside

87 "Construction of St. Anthony's Church" (photo) and "Construction of Our Lady of Guadalupe Church" (photo), Mexican-American Collection, Riverside Metropolitan Museum, Riverside, California.

88 Torres-Rouff, "Becoming Mexican," 106.

89 Zenon Sosa was among thirty-nine ethnic Mexicans from the Crestmore plant who donated money to Mexico after a dramatic earthquake. "Los Donativos de los Mexicanos en E.U. para las Víctimas del los Temblores en Mexico," *Heraldo de Mexico*, February 10, 1920, 7.

90 *Sharing Our History* (Riverside, CA: Riverside County Mexican American Historical Society, 2012).

91 Richard R. Esparza, ed. *Tales of Our Families* (Riverside, CA: Riverside Metropolitan Museum, December 31, 1995).

92 Ibid.

93 Ibid.

94 Letter from Maurice L. Oliva to Annette Sampson, "Building Permit Records Online," *City of Riverside*, March 9, 1994, laserfiche (accessed August 9, 2015).

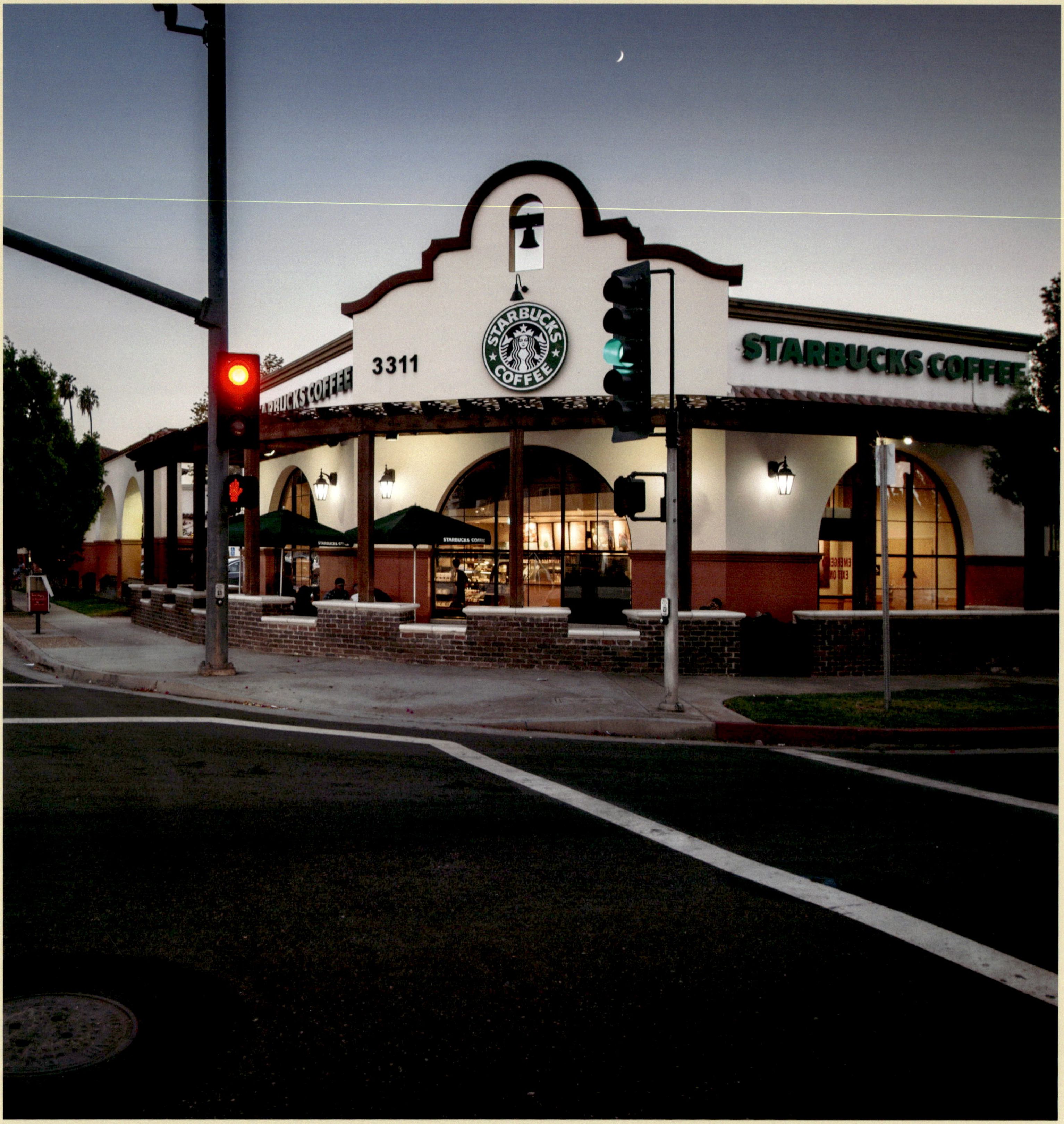
3311
STARBUCKS COFFEE
STARBUCKS COFFEE

5

Patricia A. Morton

Postwar Spanish Colonial Revival Architecture in Inland Southern California

From Mission Inn to Taco Bell

The red-tile roof, if not unique to Southern California, has become a significant regional icon. More than a mere architectural feature, red tile is visual shorthand both for the whole Spanish look of the region and for a still-resonant vision of the California good life.

—Phoebe Kropp, *California Vieja* (2006)[1]

Nothing in Spain was ever like this.

—Charles Moore, "You Have to Pay for Public Life" (1965)[2]

In recent decades, Spanish Colonial Revival architecture has entered its third renaissance. Since its earlier phases as the Mission Revival and the SCR proper, it has achieved its present apotheosis as a widespread style employed on buildings both high and low.[3] While it was a dominant style in the early twentieth-century, the SCR was little used in the immediate post–World War II period. During the 1970s, however, increasing awareness of historic, regional, and vernacular architecture, as well as the rise of postmodern architecture, produced a renewed interest in Spanish and Mediterranean styles (fig. 1).

Today, SCR architecture has become ubiquitous in the Inland Empire and in other places that celebrate the "Spanish" heritage of a fictional past. From a highbrow style patronized by a rising nouveau riche at the turn of the twentieth century, the SCR has become the lowest common denominator of the everyday environment in localities such as Riverside,

FIGURE 1 Starbucks, 3311 Market Street, Riverside, August 18, 2015

San Bernardino, Chino, Ontario, and elsewhere across the inland region of Southern California. The style has been used for postwar buildings as diverse as fast food restaurants (notably Taco Bell), mini-malls, single-family houses, big box warehouse stores, gas stations, and apartment buildings (figs. 2–5). Its omnipresence makes it a primary marker of regional identity and culture, referring to a ready-made history that any newcomer can adopt. It is also a signal of a desire for rootedness in the floating world of postwar suburbia.

Taste and SCR Architecture

During the immediate postwar period, SCR architecture played a marginal role in American taste standards. In 1949, Russell Lynes published an essay titled "Highbrow, Middlebrow, Lowbrow" in *Harper's* magazine. While it is often interpreted as a satirical send-up of American taste, Lynes' article serves as an extraordinarily clear record of postwar American taste norms. Lynes expanded this essay into his canonical book, *The Tastemakers*, in which he gave a history of taste in America beginning in the 1820s when "the long period of control over taste by a landed and intellectual aristocracy came to an end."[4] Since it was "emancipated" in the early nineteenth century, public taste has "traveled a path of windings and turnings, of peaks of frivolity and abysses of dinginess.... It has produced buildings of stark beauty and crude bombast.... But it has never been without honest striving, or without vitality or individuality."[5] The same might be said of the SCR style.

Lynes' account of taste in America began with what he called the Age of Public Taste, when taste was democratized and ceased being the exclusive concern of a few cultured people. In this period, taste reformers attempted to discipline all citizens and raise the taste of an entire nation, issuing sharp criticisms about the gauche manners and poor taste of the newly wealthy. These generalized efforts failed to instruct the taste of the American public at large, according to Lynes, and the tastemakers turned their attention to individuals who could set a better standard for the populace. Thus began the Age of Private Taste, directed primarily at the rich, who served as role models, and at middle-class homemakers, who were responsible for upholding higher ideals of refinement and culture. This Age ended in the early twentieth century with the rise of mass culture, which ushered in the Age of Corporate Taste in which mass media and big corporate entities can shepherd the taste of the masses.[6]

According to Lynes, the SCR style emerged during the Age of Private Taste as a manifestation of the "factory-produced, decorator-inspired, and inexpensive" taste of the fin-de-siècle.[7] This was the era when Gustav Stickley carried on Eastlake's moralizing tone and austere aesthetic with his Craftsman style, derived from William Morris' advocacy of handicraft, and the "Mission" furniture that set the stage for SCR's more elaborate delights. Lynes quoted Lewis Mumford who called the Craftsman bungalow and Mission furniture "the first designs that put California esthetically on the modern map" and that spread across America.[8] In this period, the humble bungalow evolved into the "Spanish Mission Style" that was the "very embodiment of homelike coziness and convenience, inexpensive but of refined elegance easily adaptable to almost any location . . ."[9] (fig. 6) Lynes saw the Spanish Mission style bungalow as the direct precursor of the postwar Ranch style house, identifying in both shallow-sloped horizontal roofs, large picture windows, and rustic details.

In a chapter based on his original *Harper's* article, Lynes explained the hierarchy of taste

FIGURE 2 Der Wiener-schnitzel, 4103 Riverside Drive, Chino, June 21, 2015

FIGURE 3 Mission Village, Market Street, Riverside, June 15, 2015

FIGURE 4 The Home Depot, 2707 S. Towne Avenue, Pomona, June 16, 2015

FIGURE 5 Shell, 3873 Pyrite Street, Glen Avon, July 6, 2015

FIGURE 6 Middle-class SCR bungalow, Castle Reagh Place, Riverside, ca. 1928, Barker Building Co., Designers and Builders

FIGURE 7 Tom Funk, "Everyday Tastes From High-Brow to Low-Brow" chart in Russell Lynes, "Highbrow, Lowbrow, Middlebrow," *Life Magazine*, April 11, 1949

in postwar America, which he maintained superseded those of class, wealth, or breeding. "Taste and high thinking" made prestige in America. "Good taste and bad taste, adventurous and timid taste, cannot be explained by wealth or education, by breeding or background."[10] Instead, a new social structure or stratification had emerged in which "the highbrows are the elite, the middlebrows are the bourgeoisie, and the lowbrows are the *hoi polloi*."[11] He illustrated his theory with a chart that placed contemporary American culture in a grid of High-Brow, Upper Middle-Brow, Lower Middle-Brow and Low-Brow, cross-indexed with clothes, furniture, useful objects, entertainment, salads, drinks, reading, sculpture, records, games, and causes (fig. 7). Whereas the High-Brow favored the ballet and Eames chairs, the Low-Brow preferred Western movies and overstuffed armchairs. This diagram represents a panorama of postwar everyday tastes frozen at a moment after World War II when Americans had the means to consume in the manner to which they felt comfortable. Instead of the moralism over bad taste that had permeated insecure mid-nineteenth-century America, Lynes identifies an ease associated with this new stratified culture.

Reconstructing Lynes' chronological diagram of American taste, SCR (in its Mission style form) would appear in the highbrow category during the 1870s-90s, quickly sliding down the taste scale to middlebrow by the 1910s–20s when it became the chosen style for suburban bungalows and service stations alike. By World War II, SCR had almost disappeared from the taste spectrum because of disuse. According to architectural critic Leon Whiteson "[b]y the late 1930s, Revival housing had become so prevalent in the Southland it was no longer chic. The upper middle-class trend setters who had first embraced Spanish architecture abandoned it as 'common.'"[12] It persisted as a lowbrow marker in older suburban subdivisions and shopping centers, but the highbrows preferred International Style modernism, "which became the avant-garde architecture of the socially and technologically hopeful post-World War II era."[13] Architecture schools ceased teaching historic styles and shifted to a curriculum based on the modernist doctrine of "form follows function" and emulated such pioneers of modernism in Southern California as Richard Neutra and Rudolph Schindler.[14]

Postwar SCR rests at the bottom of the taste spectrum. It traveled down the scale of taste hierarchy to become a broadly popular style integrated into plebian preferences. From a highbrow style associated with the elite ruling class, exemplified by the legendary Mission Inn in Riverside, it has become safe, familiar, comforting, and an alternative to more modern, challenging postwar styles. On Lynes' taste scale, SCR today has strong appeal to middlebrows and lowbrows, but less attraction for highbrows who prefer more explicitly styled architecture in modern or historic modes. As it appears on every type of building, it has no particular cachet and no distinction as a highbrow taste marker, but confers a general air of culture or history on the everyday environment (fig. 8).

Today, elements derived from the SCR are again dominant in the vernacular landscape (fig. 9). While Southern California has many architectural styles, arches, domes, red tile, and stucco dominate as the regional manner. SCR buildings first gained popularity among Anglo settlers in the late nineteenth century, evolving from Mission and Mission Revival architecture. Yet, as geographer Albert Fu notes, its ubiquity was far from guaranteed, it had to be legitimized as the regional vernacular. Fu places SCR in the context of a long effort to establish it as the preeminent California style: "going beyond its present-day banality, we see that it became a successful modern aesthetic amongst Anglos despite its historic/mythic vocabulary and Hispanic roots. At the center of Spanish-Colonial Revival is the manufacture of heritage and history."[15] Rather than evoking the period of Mexican dominion in Southern California and its hybrid people and culture, SCR fabricated a "Spanish" past that links the region to Europe and Europeans.

David Gebhard, prominent historian of California architecture, asserted that the key element of California architecture has always been "historic imagery—the suggestion of place established and maintained by allusion to the past."[16] But it was an invented history based on a romanticized and racialized understanding of regional identity. A complex view of SCR is offered by historian Phoebe Kropp who sees it as part of broader tendencies in American society: "Romantic visions of Southern California's Spanish past reflected a range of Anglo responses to modernity, from utter dismay to enthusiastic validation."[17] The construction of a Spanish heritage acceptable to Anglo settlers required effacing the immediate past of Mexican rule and excluding both former Mexicans and Native Americans from the historic narrative. This perception of the past may have been inauthentic, but Kropp sees it as "an example of a central method Americans have used to express race and nation. From blackface minstrelsy to a passion for Navajo blankets, white Americans' ability to disdain and yet desire, to reject and yet possess, was a familiar and consistent strategy for dealing with non-white people and cultures in the nineteenth and twentieth centuries."[18] The Spanish as Europeans stood for the Anglos who placed themselves at the center of the region's future while relegating all other residents to the past (fig. 10). Neither the architecture nor the way of life of the "Spanish dons" existed in reality, but images of grand *haciendas* with tiled domes, arched colonnades, and fountains in courtyards were a powerful means of crafting a regional identity for Southern California.

Postwar SCR

In inland Southern California, as globally, modernism became the dominant popular style in the postwar until a shift in the cultural zeitgeist during the 1960s and 1970s. The triumph

FIGURE 8   University Village, 1201 University Avenue, Riverside, July 8, 2015

FIGURE 9   UEI College, 1860 University Avenue, Riverside, June 16, 2015

FIGURE 10 First Church of Christ, Scientist, Riverside, November 10, 2014. Arthur Benton, architect

of modernism in commercial and residential architecture can be seen in the corporate headquarters and the mass-produced ranch house, respectively.[19] Modernism informed visions of new, technologically-advanced buildings made of new materials, such as the Monsanto House of the Future at Disneyland, designed to be built out of plastics.[20] The "Case Study House Program," sponsored by *Arts and Architecture* magazine, projected new modern houses for middle-class clients, using materials like plywood and aluminum that had been perfected in Southern California defense industries during World War II.[21] The humble tract house had a strong stylistic identity in the interwar period, but its style became more nominal in the postwar rush to build the maximum number of units as quickly and cheaply as possible.[22] Variations on a vaguely Cape Cod or Traditional style were the norm in 1950s developments in Southern California and across the U.S., and more distinct historical styles like the SCR were not congruent with the American postwar love for modern appearances. Suburbs in the immediate postwar period tended to be anonymous, standardized, and uniform, without a particular link to the place in which they were located, dictated by the builders' imperative for efficiency and speed of construction. Writing in the late 1940s, Lynes noted that "[r]egional differences in taste have all but disappeared, and if you were to be put down blindfolded in the new suburbs of any large American city it would be difficult to tell whether you were in the East or the West, the North or the South."[23] While Levittown might be the most famous postwar American suburb, Southern California had its share of mass-produced suburbs, such as Westchester and Panorama City. Inland, former streetcar suburbs like Pomona and self-contained cities like Riverside ballooned with starter homes and shopping strips modeled after those found nationwide.

One exception to this tendency to more abstract architecture was the Ranch style house, derived from reference to Spanish haciendas or pioneer farm buildings but lacking explicitly SCR details. The Ranch style had its origins in California and found wide acceptance as an alternative to the standard Cape Cod type that was built from Levittown, New York, to Lakewood, California. Gebhard notes the Ranch style's ubiquity: "The post-World War II California Ranch house of Cliff May, exposed by *Sunset* magazine, was a sought-after idea which pervaded the whole of the country."[24] May, the most famous practitioner of the Ranch style, claimed that he had been inspired by the adobe houses of the early California rancheros and by the "whole California way of living." But while the Ranch style house incorporated some elements of the SCR house, it avoided the historic models that had been so popular in the early twentieth century. According to architectural historian Merry Ovnick, "clay tile roofs became the mark of houses no longer new. Arches were incompatible with the vernacular form of the Ranch style house."[25] The postwar Ranch house was an amalgam of modern architecture and vernacular style, with its low, horizontal roof, minimally rustic decorative details, picture windows and inside-outside floor plan (fig. 11). Like SCR style, it was born out of a romantic vision of casual living and the good life in California, what Lynes called "the pervasive Western spirit of the open range and the barbecue, of sunshine and leisure."[26] The Ranch House had no specific regional associations, however, and could substitute for the Cape Cod or modern style in any locale.

In the 1960s, the postwar ethos of optimism about the future and faith in American capitalism began to fade. Prosperity and progress were the bywords of the immediate postwar era, allied with a faith in experts and expertise, and expressed in a fundamental confidence about the future. The cultural and political turmoil of the 1960s, however, diminished this optimism and brought about a distrust of elites, a search for alternative cultures, an embrace of popular culture, and a renewed interest in history.[27] Rather than drawing on a codified set of forms established by modernist "masters," architecture in the '60s and '70s moved toward eclecticism derived from history and contemporary culture, including vernacular, marginal, and popular culture. In this period, SCR architecture enjoyed a revival as a style that seemed to evoke a more cultured, rooted period with its characteristic tile roofs, stucco walls, arches, and domes (fig. 12). By the time British architectural historian Reyner Banham wrote his iconic book, *Los Angeles: The Architecture of Four Ecologies*, in the early 1970s, SCR architecture was again so pervasive as to become invisible, worthy of no particular attention. Banham went so far as to declare that SCR was not "an identifiable or consciously adopted style" but "something which is ever-present and can be taken for granted, like the weather—worth comment when it is outstandingly beautiful or conspicuously horrible, but otherwise simply part of the day-to-day climate...."[28] He referred to both the Spanish-inspired architecture of the early twentieth century and the later "Mediterranean" mode of the interwar period with references to Mexican, Spanish, Italian, or Moorish architecture.

The re-revival of SCR style had a number of cultural and social motivations. In architecture, the 1970s were a period of renewed interest in historical styles, culminating in the emergence of postmodernism. Synthetic or historicist, postmodernism in architecture attempted to supersede the abstract language of modernism and restore meaningful form that resonated with users.[29] Accompanying this return to historical architecture, a revived concern for local styles, rather than the universal mode of modernism, stimulated new theories of critical or radical regionalism.[30] In an era of increased mobility among Americans, the faceless, anodyne suburbs of postwar sprawl seemed to have less and less connection to place, a quality that had heightened importance for American culture. Architect Charles Moore and others called on architects to recover the symbolic function of design and create orderly places out

FIGURE 11 House, Phoenix Place and Deerfield Street, Ontario, June 21, 2015

FIGURE 12 Rite Aid, 3849 Chicago Avenue, Riverside, August 18, 2015

of the "chaos" of the postwar environment. As Moore put it, "We are in urgent need of understanding *places* before we lose them, of learning how to see them and to take possession of them."[31] Moore designed such buildings as the Anawalt house in Point Dume (1988) as an exercise in making place out of the SCR, which he called "a Southland archetype . . . the image of our transformed semi-desert, climatically Mediterranean landscape, the architecture of our innocence."[32] After the turn from modern architecture's universal ideal, which erased any local identity from a building's appearance, the ornamented surfaces and faux-historical references of SCR began to gain appreciation and to seem congruent with postmodernism and regionalism alike.

In the 1970s and '80s, builders began offering more choices than the simplified Cape Cod or Ranch style units that predominated in the immediate postwar suburbs. The standardized, mass-produced character of the postwar built environment became a marketing handicap, rather than a selling point. In his study of "suburban aesthetics," architectural historian John Archer described how historian Adrian Forty and architect Henry Moss analyzed the provision of "quasi-vernacular and quasi-historical style choices which, as marketed to suburban house buyers, not only attempt to camouflage the standardized and mass-produced nature of suburban tract housing, but also fashion an ostensible 'scenery of permanence' and myth of authenticity for those who live there."[33]

In Southern California, the desire for a more rooted existence and an architectural style to match it found expression in a revival of the region's previously favorite mode: SCR. As Leon Whiteson noted in 1989, the persistent appeal of the Spanish Revival Style: attracted a wide spectrum of designers, from architects who work within the mass market residential subdivision industry to those who plan large private villas. . . . The city of Irvine is awash in pink stucco and red tiled roofs. The rapidly developing San Gabriel Valley is populated with new speculative housing tracts designed in some variation of the Revival style.[34]

The new SCR provided an instant regional identity to the masses that moved to the Inland Empire from elsewhere, particularly from Los Angeles and Orange County where real estate was more expensive. In search of more affordable housing, the new arrivals moved into environments that simulated more affluent places where SCR architecture was part of the brand: Santa Barbara, San Juan Capistrano, or Rancho Santa Fe. As did their wealthier antecedents, the names of the new suburban communities evoked the "Spanish" past, such as Hacienda Heights or Rio Vista. In a manner similar to early twentieth-century Anglos, today's migrants to the Inland Empire pursue a heritage with which to stake a claim in new territory.

FIGURE 13   University Village, 1201 University Avenue, Riverside, July 8, 2015

FIGURE 14 First Taco Bell Restaurant, Downey

What should the new SCR be called? In 1990, critic Lawrence Cheek investigated the emergence of a third Spanish revival in the Southwest and uncovered a number of neologisms. He rejected "Spanish Colonial Revival Revival" because it read like a typing mistake. Architects made clear their disdain for this style by dubbing it "Taco Deco" and "Mariachi Moderne" (fig. 13). While developers and builders settled on the diffuse associations of "Southwestern/Mediterranean" or "Spanish," Cheek himself preferred "refried architecture."[35] Other options include Neo-Spanish Colonial Revival, Southwestern, and California Style, the latter of which evokes the strong affiliation of this style with a casual, outdoors-oriented lifestyle. "Spanish" has become the default signifier for buildings with tile roofs and stucco walls. One subdivision, Estancia in Eastvale near Chino, for example, advertises "Old World Style. New World Luxury" in "traditional Spanish architecture."[36]

The "refried" SCR could be applied to any postwar suburban building type without reference to its use. It decorates tract houses, mini-malls and shopping centers, churches, liquor stores, health clinics, office buildings, gas stations, civic structures, motels, big box retail, and fast food outlets. The Taco Bell restaurants are one of the most famous manifestations of postwar neo-SCR. In 1962, Glen Bell, a restaurant entrepreneur, opened the first Taco Bell in Downey, which formed the model for a standardized design with red-tile roofed, triple-arched porch, *espadaña* gable, and *vigas* protruding on the side façades (fig. 14). Bell wanted his restaurants to look as authentic as possible, to convey the impression that they sold authentic Mexican food, therefore he built them out of slump stone and tan brick to resemble an adobe house.[37] His vision included shops and other food stands lining the plaza, flanked with fire pits and a stage in the back for live mariachi performances.[38] The Taco Bell *espadaña* gable façade is instantly recognizable. While Bell stopped using the original design because it did not support a drive-thru, the characteristic gable end spread SCR across the country.[39] The prototype Taco Deco, the Taco Bell style is a hybrid, "Mexican-style" variation on Mexican architecture, just as Enchiritos, Gorditas, and Crunchwraps are variations on classic Mexican food. It has become part of Southern California's postwar zeitgeist and mythmaking. The first Taco Bell restaurant, "Numero Uno," was decommissioned in 1986, but it has become a landmark worthy of preservation. In November 2015, the structure, which had been vacant and was threatened with demolition, was moved to the Taco Bell headquarters in Irvine.[40]

The Postwar Spanish Colonial Revival Lexicon

Like its predecessors the Mission Revival and the SCR, neo-SCR architecture relies on a few simple components to signal its "Spanishness." David Gebhard identified this tendency as an early characteristic of the style from its first appearance in California:

> Neither the essential forms nor the structure of the Mission Revival buildings had anything to do with their supposed prototypes. Instead, the Mission Revival architects conjured up the vision of the Mission by relying on a few suggestive details: simple arcades; parapeted, scalloped gable ends (often with a quatrefoil window); tiled roofs; bell towers (composed of a series of receding squares, normally topped by a low dome); and finally (and most important), broad, unbroken exterior surfaces of rough cement stucco.[41]

Gebhard's description provides a lexicon of elements that remain the mainstays of neo-SCR. What Lawrence Cheek refers to as "mission mascara" can be applied to a standard building type of conventional construction in the form of an *espadaña*, a tile roof, arches, a cross, or a dome (fig. 15).[42] While the details of early twentieth-century SCR buildings were often elaborate and crafted, the features of refried SCR have a sketchy, minimalist quality like the applied moldings on Robert Venturi's house for his mother, without the irony.[43]

The dominant mode of building today in Southern California, balloon-frame construction, allows a structure's façade to take on the characteristics of the style, thereby creating a disjunction between interior and exterior. Form and function are disconnected. The shed, box, or hut can be dressed in a fancy suit of SCR elements, creating what anthropologists Alessandro Falassi and Edward Tuttle call "an architecture of costume."[44] A mask

FIGURE 15 CVS Pharmacy, 3361 Market Street, Riverside, June 15, 2015

of historical allusions disguises the banal architecture of everyday functions, vaguely connoting fictional narratives of "Spanish" California or another exotic locale and period. The operation might result in total masquerade or a more complex interaction between representation and function.[45]

Each of the elements in the lexicon serves as an interchangeable signifier of SCR that can be applied to any anonymous shell. The basic lexicon consists of roof tile, stucco, arch, dome, and *espadaña*, which can be supplemented by columns, buttresses, wrought iron, slumpstone or rough brick, *campanario,* and other details. What follows is an outline of the SCR lexicon in text and photographs.

Tile Roof

The tile roof is the most ubiquitous and important element in the SCR lexicon (fig. 16). Tiles are so strongly associated with SCR that a recent study of Spanish Revival architecture in the United States is titled *Red Tile Style*.[46] In addition, the tile roof is a deep symbol of a California lifestyle that promises to be accessible to everyone. The expanses of red tile that cover homes, offices, shopping malls, and service stations offer visual testimony to Southern California's investment in the Spanish-inspired good life. As Phoebe Kropp notes, "this distinctive regional cap seems to have proliferated in recent decades, even as public faith in the availability of the good life wanes and Southern California appears to be tragically bereft of its former magic in the popular imagination."[47] A few lines of red tiles along a roofline instantly transform a strip mall or a gas station with a bit of SCR graciousness (fig. 17). By encrusting buildings with tile, other parts can be standard builder's components with no connotative power and still convey a refried SCR style.

Stucco

Concrete stucco is a more neutral building material than the red tile roof, and it must generally be supplemented with other decorative elements to qualify as neo-SCR (fig. 18). Cement stucco was an essential part of early twentieth-century costumed architecture because it could be used over a wood balloon frame to produce almost any style, but it had a particular affinity for SCR buildings. Gebhard quotes Paul Edgar Murphy who wrote in 1928: "If a house has its exterior covered with stucco, it is Spanish."[48] As Fu has shown, in reference to the original SCR style of the 1910s and 1920s, cement was sold as an affordable and durable material that resisted the effects of time and nature and protected against decay, as opposed to the more authentic Mexican adobe, which signified decadence and inferiority in the form of the ruined missions.[49] As a modern material, stucco could provide more permanence complementing its adaptability to any style. Color plays an important role in allowing stucco to conjure a properly SCR ambiance. Shades of pink, ochre, and orange make up the basic SCR palette, but more outré tones of purple, red, and green have a place in this style (fig. 19).

Dome

The dome is largely reserved for civic and commercial buildings, and is rarely found in the more humble forms of domestic neo-SCR architecture (fig. 20). It connotes grandeur and monumentality as evidenced by the Mission bell towers crowned by domes. At the turn of the twentieth century, Arthur Benton designed a domed tower at the corner of Riverside's Mission Inn that has become the archetype for countless domes across the Inland Empire. Often capping a *campanario*-style tower, refried SCR domes provide landmark profiles to otherwise banal structures (fig. 21).

Arch

Arches and arcades signal SCR or Mediterranean architecture immediately, countering the orthogonal regularity of wood frame construction with great semiotic efficiency (fig. 22). Current Taco Bell locations feature a highly stylized arch surmounting a bell logo, the only remaining fragments of the original Taco Bell model. Arches are the lowest common denominator element of neo-SCR, the most easily applied to an anonymous shell and capable of being simulated in stucco without more costly embellishment (fig. 23). Supported on columns of a indeterminate origin, arches form rudimentary arcades that add rhythm and shaded walkways to strip malls and big box retail stores, as well as porches on houses.

Espadaña

The *espadaña* gable is one of the most iconic elements of neo-SCR style (fig. 24). Derived from bell gables found on churches throughout Spain and former Spanish colonies, the *espadaña* substituted for a church bell tower, being cheaper and easier to build.[50] In postwar SCR examples, the stepped form and arched niches may or may not hold bells. More than any other element of the SCR lexicon, the *espadaña* creates a strong association with "Spanishness" even when tripped of all detail (fig. 25). It appeared on the first Taco Bell restaurant, making it one of the most highly disseminated elements of SCR. Roof parapets in a variety of curved and stepped shapes evoke the *espadaña* as found on the mission prototypes, but deviate from the precedent in detail and form.[51]

Conclusion

A counter to the anonymity and alienation of contemporary suburban sprawl environments, neo-SCR gives the appearance of rootedness and history to the region's postwar architecture.

FIGURE 16 Village Liquor, 4117 Riverside Drive, Chino, June 16, 2015

FIGURE 17 Alberto's/Circle K, 2727 S. Reservoir Street, Pomona, June 16, 2015

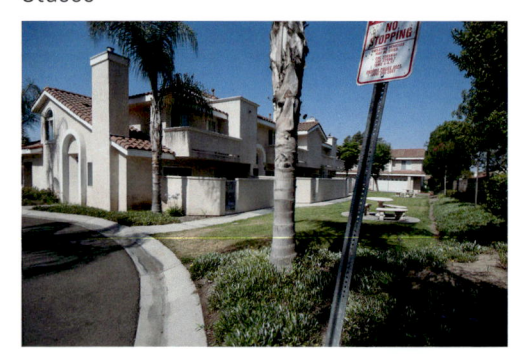

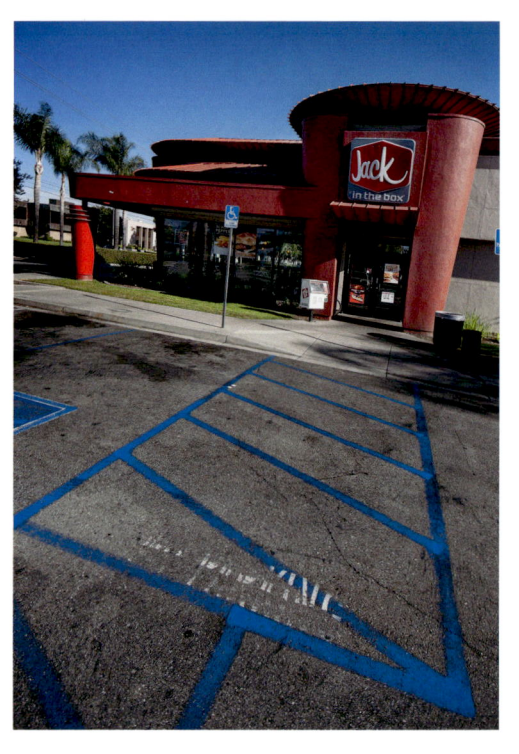

FIGURE 18 Condominiums, 2900 S. Campus Avenue, Ontario, June 21, 2015

FIGURE 19 Jack in the Box, 2775 S. Reservoir Street, Pomona, June 16, 2015

Dome

Arch

Espadaña

FIGURE 20 Raincross Promenade, 3250 Market Street, Riverside, August 18, 2015

FIGURE 21 Pomona Marketplace, 2700 S. Towne Avenue, Pomona, June 16, 2015

FIGURE 22 DaVita Pomona Dialysis Center, 2703 S. Towne Avenue, Pomona, June 16, 2015

FIGURE 23 City of Riverside Garage Seven, 3601 and 3605 Market Street, Riverside, June 15, 2015

FIGURE 24 CVS Pharmacy, 3361 Market Street, Riverside, June 15, 2015

FIGURE 25 Toys R Us (closed), 2727 S. Towne Avenue, Pomona, June 16, 2015

While never an authentic or sincere style, revived SCR architecture creates a sense of cultural identity in a fragmented, chaotic locale and connects the present to a mythical past. Synthetic memories of a more harmonious, coherent time, whether the fictional period of the Spanish dons or the subsequent golden age of Anglicized California, adhere to the barest details of contemporary SCR buildings. What Douglas McCulloh calls the "mutant offspring" of the original SCR monuments serve as cultural signposts in a region where most residents have no long-term attachment to the place (fig. 26).[52] Its ubiquity is its greatest strength and its greatest limitation. According to Kropp, "The trouble with the red-tile roof is that it appears to be merely an ornamental commodity, but it is more important than that. Its instinctive, habitual consumption elevates rather than diminishes its cultural influence. As a popular abbreviation for the region's Spanish character, it both inspires remembrance and obscures it."[53] The memory of the first Spanish settlers in the area, who migrated from Arizona, is effaced in favor of the romanticization and stereotyping produced by SCR buildings and culture. Elizabeth McMillian has suggested that "Spanish colonial revival may

FIGURE 26   Office Max (closed), 2700 S. Towne Avenue, Pomona, June 16, 2015

speak more persuasively to the population of the Los Angeles area in the 1990s than it ever really did in the 1920s," indicating that the formerly elite, highbrow style of the early twentieth century has found its widest audience among the masses living on the edges of today's Southern California.[54] A marker of lowbrow taste, refried SCR architecture appeals to a broader public, while it promulgates the fabricated history of a domesticated "Spanish" past. Simultaneously alive and degenerate, neo-SCR imparts a regional flavor to the otherwise bland Inland postwar environment and evokes a still-potent myth of the California good life.

Notes

1. Phoebe Kropp, *California Vieja: Culture and Memory in a Modern American Place* (Berkeley: University of California Press, 2006), 261.
2. Charles Moore, "You Have to Pay for the Public Life," *Perspecta* 9–10 (1965): 60.
3. In his canonical article on the SCR, David Gebhard gives a rough chronology for the first two phases of the style. According to Gebhard, the Mission Revival began in the 1880s and "reached its fullest development during the first decade of the twentieth century." He dates the high point of the SCR between 1910 and the early 1930s. David Gebhard, "The Spanish Colonial Revival in Southern California (1895–1930)," *Journal of the Society of Architectural Historians* 26, no. 2 (May 1967): 131–32.
4. Russell Lynes, *The Tastemakers* (New York: Grosset & Dunlap, 1954), 5.
5. Ibid., 7.
6. Ibid., 6–7. The present moment appears to be an extension of the Age of Corporate Taste, although it might be interpreted as a return to the period of Private Taste and its extraordinary wealth and taste inequality.
7. Ibid., 186.
8. Ibid., 188.
9. Ibid., 189.
10. Ibid., 310.
11. Ibid.
12. Leon Whiteson, "'20s Spanish Style Needs No Revival. Red Tile Roofs, White Walls Endure as Southland Favorites," *Los Angeles Times*, March 5, 1989, http://articles.latimes.com/1989-03-05/realestate/re-402_1_spanish-style (accessed July 28, 2015).
13. Ibid.
14. Among the many sources on modernism in Southern California, see Esther McCoy, *Five California Architects*, (New York: Reinhold, 1960); and Thomas S. Hines, *Architecture of the Sun: Los Angeles Modernism, 1900–1970* (New York: Rizzoli Press, 2010).
15. Albert S. Fu, "Materializing Spanish-Colonial Revival Architecture: History And Cultural Production In Southern California," *Home Cultures* 9, no. 2 (July 2012): 151–52.
16. David Gebhard, "Architectural Imagery, the Mission and California," *The Harvard Architecture Review* 1 (Spring 1980): 145.
17. Kropp, *California Vieja*, 4.
18. Ibid., 7.
19. On the corporate headquarters, see Reinhold Martin, *The Organizational Complex: Architecture, Media, and Corporate Space* (Cambridge, MA: MIT Press, 2003). On the mass-produced tract house, see Barbara Miller Lane, *Houses for a New World: Builders and Buyers in American Suburbs, 1945–1965* (Princeton: Princeton University Press, 2015).

20 See Stephen Phillips, "Plastics," in *Cold War Hothouses: Inventing Postwar Culture, from Cockpit to Playboy*, ed. Beatriz Colomina, Annmarie Brennan, and Jeannie Kim (New York: Princeton Architectural Press, 2004), 91–123.

21 See Elizabeth A. T. Smith, Julius Shulman, and Peter Goessel, eds. *Case Study Houses* (New York and Köln: Taschen, 2002).

22 See Greg Hise, *Magnetic Los Angeles: Planning the Twentieth-Century Metropolis* (Baltimore and London: Johns Hopkins University Press, 1997), 56–85.

23 Lynes, *The Tastemakers*, 254.

24 Gebhard, "Architectural Imagery, the Mission and California," 145.

25 Ibid., 291.

26 Lynes, *The Tastemakers*, 254.

27 Alice T. Friedman, *American Glamour and the Evolution of Modern Architecture* (New Haven and London: Yale University Press, 2010), 226–27.

28 Reyner Banham, *Los Angeles: The Architecture of Four Ecologies* (New York: Penguin, 1971), 61.

29 See Charles Jencks, *The Language of Post-Modern Architecture* (New York: Rizzoli, 1977); Paolo Portoghesi, *Postmodern, the Architecture of the Post-Industrial Society* (New York: Rizzoli, 1983); Heinrich Klotz, *The History of Postmodern Architecture* (Cambridge, MA: MIT Press, 1988); and Jorge Otero-Pailos, *Architecture's Historical Turn: Phenomenology and the Rise of the Postmodern* (Minneapolis and London: University of Minnesota Press, 2010).

30 Most notably, Kenneth Frampton espoused Critical Regionalism as a method for overcoming the anonymity of International Style architecture and recuperating aspects of regional culture while rejecting the willfulness and scenography of postmodernism. Kenneth Frampton, "Towards a Critical Regionalism: Six Points for an Architecture of Resistance" in *The Anti-Aesthetic. Essays on Postmodern Culture*, ed. Hal Foster (Seattle: Bay Press, 1983), 16–30.

31 Donlyn Lyndon, Charles W. Moore, Patrick J. Quinn, and Sim van Der Ryn, "Toward Making Places," *Landscape* 12, no. 1 (Autumn 1962): 34.

32 Quoted in Whiteson, "20s Spanish Style."

33 John Archer, "Suburban Aesthetics Is Not an Oxymoron," in *Worlds Away: New Suburban Landscapes*, Andrew Blauvelt, ed. (Minneapolis, Minn.: Walker Art Center, 2008), 138. See Adrian Forty and Henry Moss, "The Success of Pseudo-Vernacular," *Architectural Review* 167, no. 996 (February 1980): 73–78.

34 Whiteson, "20s Spanish Style."

35 Lawrence Cheek, "Taco Deco: Spanish Revival Revived," *Journal of the Southwest* 32, no. 4 (Winter 1990): 491–98.

36 http://www.lennar.com/new-homes/california/inland-empire/eastvale/estancia (accessed December 29, 2015).

37 Nancy Luna, "Taco Bell Turns 50," *The Orange County Register*, March 21, 2012, http://www.ocregister.com/articles/bell-345567-taco-turns.html (accessed September 9, 2014).

38 "Save Taco Bell Numero Uno," https://www.tacobell.com/feed/savetacobell (accessed January 1, 2015).

39 Luna, "Taco Bell Turns 50."

40 Sarah Parvini, "Adios, Taco Bell: Original store moves from Downey to Irvine in late-night run," *Los Angeles Times*, November 20, 2015, http://www.latimes.com/local/lanow/la-me-ln-original-taco-bell-move-20151120-story.html (accessed December 31, 2015).

41 Gebhard, "The Spanish Colonial Revival," 134.

42 Cheek, "Taco Deco," 495.

43 See Stanislaus von Moos, *Venturi, Rauch & Scott Brown: Buildings and Projects* (New York: Rizzoli International Publications, 1987), 244–46.
44 See Alessandro Falassi and Edward Tuttle, "California's Houses in Costume," *The Journal of American Folklore* 103, no. 410 (October–December 1990): 503.
45 Ibid.
46 Arrol Gellner and Douglas Keister, *Red Tile Style: America's Spanish Revival Architecture* (New York: Viking Studio/Penguin Group, 2002).
47 Kropp, *California Vieja*, 261.
48 David Gebhard, "Tile, Stucco Walls and Arches: The Spanish Tradition in the Popular American House," in *Home Sweet Home: American Domestic Vernacular Architecture*, ed. Charles W. Moore, Kathryn Smith, and Peter Becker (New York: Rizzoli, 1983), 104.
49 Fu, "Materializing Spanish-Colonial Revival," 162.
50 See, for example, Richard Perry and Rosalind Perry, *Maya Missions: Exploring the Spanish Colonial Churches of Yucatan* (Santa Barbara: Espadaña Press, 1988).
51 See Abby Moor, "Eclectic Revivals," *The Houses We Live In: An Identification Guide to the History and Style of American Domestic Architecture*, ed. Jeffery Howe, (San Diego: Thunder Bay Press, 2002), 310.
52 Email to author, August 28, 2015.
53 Kropp, *California Vieja*, 269.
54 Elizabeth McMillian, *Casa California: Spanish-Style Houses from Santa Barbara to San Clemente* (New York: Rizzoli, 1996), 19.

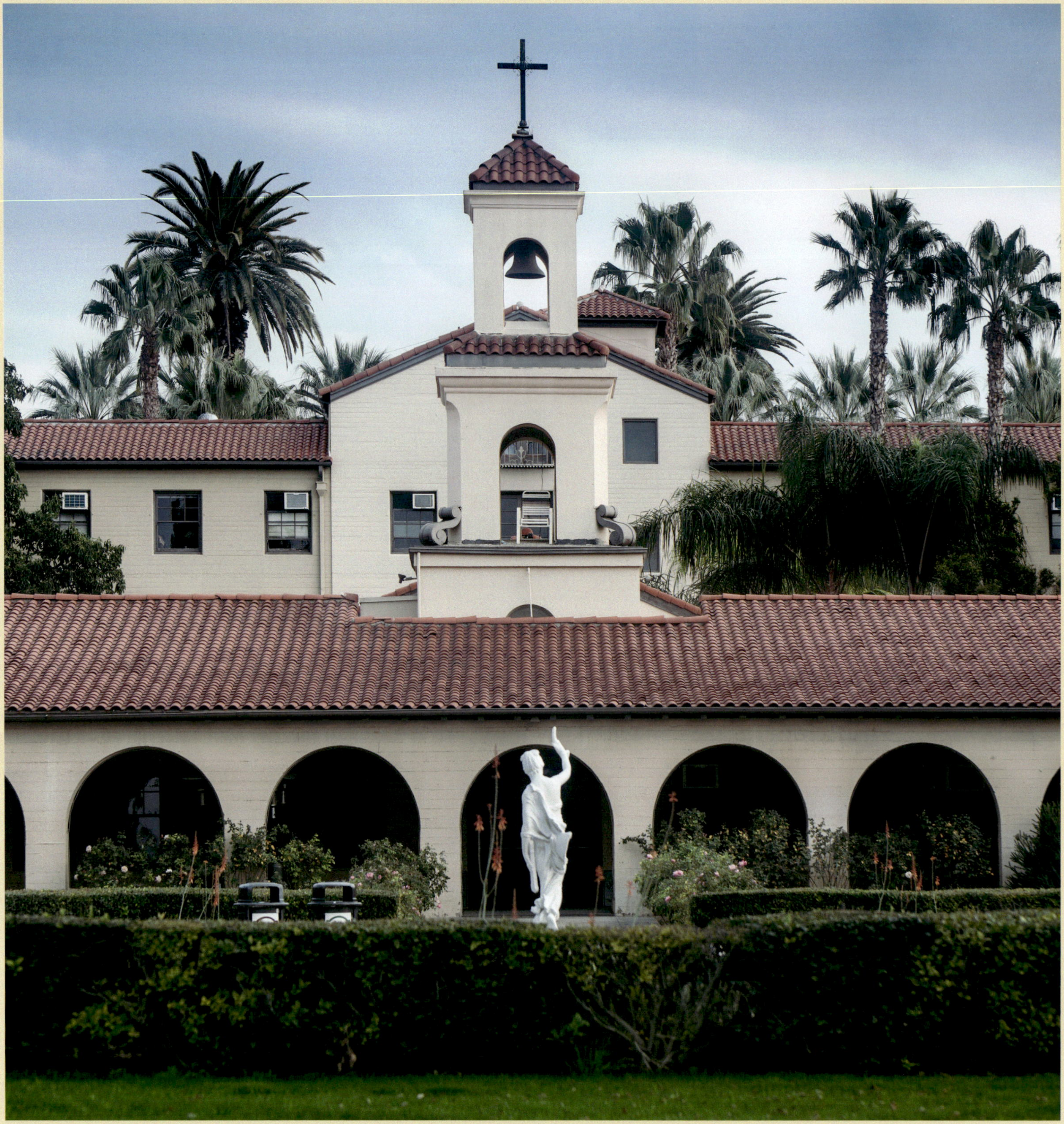

6

H. Vincent Moses and Catherine Whitmore

The Spanish Colonial Revival at California Baptist University

Interview with Dr. Ron Ellis, President

Founded originally as California Baptist College, the dynamic campus of California Baptist University (CBU) stands near the geographic center of modern Riverside at 8432 Magnolia Avenue. Twenty-two years ago, it wasn't so dynamic. By 1994, the school was a small financially struggling Christian liberal arts campus, looking for the right leader to flip its fortune. Dr. Ron Ellis proved to be the man, and the board hired him as President in 1994 with a mandate to lead the college to a brighter future.

Under Ellis' administration, the turnaround came rapidly. CBC quickly advanced to full-fledged university status as CBU, experiencing a remarkable reversal of fortune, and an exponential expansion of its schools and colleges, with an accompanying building program to match.

Architecture played a major role in the success of Ellis' plan. The University's original buildings were designed in a distilled Spanish Colonial Revival style by noted local architect, Henry L. A. Jekel, ca. 1925–38, as a retirement home for the Neighbors of Woodcraft (fig. 1). President Ellis saw these romantic Spanish Revival buildings as key to the future architectural theme for the school. In the midst of transformative growth, Ellis has maintained an unswerving commitment to the preservation of Jekel's original buildings, and to a tactical application of the Spanish Revival in the design and construction of new campus buildings.

Dr. Ellis sat down with H. Vincent Moses and Catherine Whitmore on September 26, 2016, to discuss how he came to Riverside, his

FIGURE 1 Detail of west elevation, James Complex, CBU, 2016. Henry L. A. Jekel, architect

mission-driven turnaround of the institution, and his rationale for the application of Jekel's style as a model for the campus building program (fig. 2).

Moses Thank you, Dr. Ellis, for speaking with us. For the record, please tell us about your educational background and what motivated you to take on the monumental task of turning around a small financially struggling Christian college?

Ellis I earned a BA from Houston Baptist University and a Master's from Baylor University in Educational Administration, with a minor in Church–State Studies. While at Baylor, I really felt called to be a president of a Baptist college, particularly one that needed to be turned around. In preparation for my calling, I began to study organizational climate, especially organizational change—how does one initiate systemic change without causing damage to the organization? From Baylor I went immediately to Texas A&M for my PhD studies and residency and continued to delve deeply into organizational change theory.

Moses How did you get from Texas to Riverside and CBC?

Ellis I was the Executive Vice President at Campbellsville College in rural Kentucky, which is a Baptist institution. There I met Allen Wayne, a retired school superintendent, and his wife Dottie, from California, where he had been a trustee at CBC in the early 1980s. When Russell Tuck retired as CBC president in 1994, Wayne contacted me to say that he wanted to recommend me for the position.

The next Thursday I was contacted by Bill Hogue, the Executive Director of the California Southern Baptist Convention, who heard I might be interested in the job. We spent at least forty-five minutes in conversation about the school and the position. Over the next few months I spent at least 150 hours researching the college and the Inland Empire as well as the higher education environment. From that I formulated a turnaround plan to take with me to the interview.

I spent my master's and doctoral programs studying the theory of how to turn around religious colleges and similar organizations, concluding that I now saw it as sort of a calling. CBC presented itself as an opportunity to put it all together, so I was interested.

When I started reading about this area called the Inland Empire, which I had never heard of, I soon became fascinated with the large population and the potential that was here, because these were not present in the rural areas where I had worked. I saw a tremendous opportunity to serve and minister through the job at CBC. My research showed that the Inland Empire was projected to rapidly grow, but that it trailed other successful regions in many infrastructure indicators. With that in

FIGURE 2   CBU President Dr. Ron Ellis standing in north portico, James Complex, California Baptist University (CBU), Riverside, January 2017. Henry L. A. Jekel, architect

mind, I put a plan together on how CBC could meet a number of those educational needs and stay true to its religious calling. My plan was based in part on the mission of the historic Land Grant Colleges, created by the Morrill Act of 1862. Most of the Christian colleges tend to be small liberal arts colleges; but what if they were a "go to" place for practical applications as well, meeting the needs of the society? The Board of Trustees liked that idea, taking a chance on a thirty-eight-year old, and my wife and I moved with our eleven- and nine-year old sons in 1994.

Moses What was your first impression when you arrived on campus? What impact did the James Complex (fig. 3) and its architectural style have on you?

Ellis When we came out to CBC the first time for my interview, I was looking for a sign that this was the right place. Since the age of nine, the Great Commission, Matthew 28:19–20, where Jesus gave his remaining apostles their marching orders to take the gospel to the world, has been the very core of my Christian walk. It was my desire to lead a university committed to the Great Commission and CBC seemed like a great fit.

Jane and I were housed at the Mission Inn for that interview. A CBC employee drove us to the campus and dropped us off in front of the James Complex, near the Fortuna Fountain. As we approached the James building and just before we entered those massive wooden doors, I pointed to a foundation stone and said, "Look at that!" It was Matthew 28:19–20, and it was in stone. I felt a chill down my spine, and thought, "wow, this could be the place!" I cannot emphasize how powerful that feeling was as we entered the building.

As far as the architecture goes, those massive wooden entry doors, bell tower, and the arches really created a sense of place; enhanced by that big lawn across the front between the James Complex and Magnolia Avenue. We came up Palm Drive, the main entry road back then, and made that little turn by the Fortuna Fountain, where we really experienced a sense of arrival. You knew you were some place, not just at an exit or stop along the road; I had a sense that we were at a special place.

After a tour of the campus, we met in the Great Hall (Staples Room) in the James Building. That room has a strong Spanish Colonial Revival character, especially in tandem with the Sun Parlor of wood and art glass adjacent to the Hall. It was impressive, despite deferred maintenance. I could see the potential of that room, and the feel of a "great good space."

Moses What in your personal experience growing up or in your training provided you with this sensitivity to the role of architecture in creating a "great good space?"

Ellis I was born in Natchitoches, Louisiana, founded in 1714 in the northern part of the state. The town has a distinctive French heritage and it is almost like taking a part of the New Orleans French Quarter and transporting it to the other part of the state. I particularly remembered as a child listening to the thump-thump-thump of the car tires as they drove over the brick pavers. That sound stuck with me. It just kind of filled the senses; somehow it was comforting, and I never forgot that. In 2003 when we built the Yeager Center and created a new entrance to the campus with a new bridge, I had pavers installed. I wanted to make sure those pavers had that sound, thump-thump-thump, as people drove into the university, a welcoming and comforting experience.

In the 1960s, from the age of five to thirteen, I lived in Las Vegas with my family. One of the things I remember about that time was the non-stop construction and a sense of imagination and fantasy in that construction. It was not unusual for visionaries to design something like Caesar's palace and then build it in the Nevada desert. So even as a young child, that was an impression that stayed with me concerning the power of architecture and how you can create a sense of place and transport people virtually to another time and place.

In the late 1980s I worked in Pat Neff Hall at Baylor University. This Hall was the main administration building for Baylor and a very impressive four-story Georgian Revival structure. I was taken by how they renovated the interior but left the exterior intact. The interior of the building was brought into the modern era, with updated comforts and technology, yet they left the exterior as timeless as it had been since construction in the 1930s. That was another one of those things I filed away.

Moses Given what you just said, how did that play a role in the way you envisioned your turn-around plan for CBC?

Ellis I have probably visited over a thousand colleges and universities around the world. My wife knows that when we are traveling if there is a campus within a few miles, I am at least driving through it or if I have time, I am probably walking some of it, just to experience how it feels and what it says to me. How does it speak to me and does it have features about it that have been done well that I can file for future use? I was always impressed with college campuses that had an obvious architectural theme and stayed with it. My idea was that

FIGURE 3 Front elevation, James Complex, CBU, built as the Neighbors of Woodcraft Home, ca. 1925–38. Henry L. A. Jekel, architect

when I became a president at a college, whether that style was Cape Cod, Georgian, or Spanish Colonial Revival, we were going to stick with it. When I came here for the first time and saw how extensive the Spanish Colonial Revival buildings were on campus, the very core of the campus, I thought "wow," this would be an amazing signature to retain for future expansion.

Moses Henry L. A. Jekel, the architect of your original buildings, designed them in a distilled SCR style, vaguely derived from the California Missions, and set far back from Magnolia Avenue to emulate the pastoral setting of a mission. What specifically about Jekel's design and layout appealed to you, and how did it play into your new plan for campus expansion?

Ellis One of the things I noticed when I arrived was the beautiful expansive front lawn with Palm Drive leading to the front of the James Complex. It created an obvious sense of arrival and that was something I wanted to emphasize about CBC. I wanted to give people a sense of the pastoral setting as they drove into the campus, with its inherent ability to produce a feeling of relaxation. I wanted them to feel like they had arrived and left all the hectic things outside the gates. We host a number of prospective students and first-time visitors, so those first impressions are important. Likewise, the feeling of arriving in a pastoral setting is comforting regardless of the number of times one enters the CBU campus.

To help us achieve this feeling we enclosed the front of the campus in a wrought-iron fence to confirm that you have arrived. In my mind, it was important that a fence defines the front of the campus with the expansive lawn behind it, often with students playing soccer or throwing a Frisbee. This is a very pleasing view even if you just drive by the campus.

We continued the theme with the creation of the new entrance off Magnolia Avenue, which came with the completion of the Yeager building in 2003. It includes the massive entrance monuments, large palm trees, attractive foliage, and an inviting kiosk, along with the thump-thump-thump sensations while crossing the bridge and approaching the Yeager Center (fig. 4).

Moses Along with the pastoral setting, what was it about the character of Jekel's Spanish Colonial Revival buildings that appealed to you and played into your expansion plans?

Ellis A sense of permanence. I visited Oxford University in July 2016, probably my seventh or eighth time. Oxford always exudes that sense of timelessness that a university ought to have. The architecture at Oxford elevates the human spirit; it says you are in a place of learning, beauty, and goodness. I felt the same way about Jekel's Spanish Colonial Revival buildings at CBC when I first saw them. I tried to apply the same concept in the new buildings and layout of our expanding campus. When one walks around the buildings, whether it is the original James Complex or a brand new one, there is something of visual interest in it. For instance, we have incorporated several characteristics from the original buildings, such as the arches, extensive wrought iron, and the finer details that one sees on the buildings.

Most notable to me is that everything about the original James Complex is understated. It is not baroque or rococo or overdone. It has very simple lines, massive in their design, and it also incorporates art tile and dark wood doors and beams, as well as other features that we have sought to emulate in the new buildings. One technique of note about the James building is that when they poured the concrete they used the same method as used for the

International Wing of the Mission Inn and other Spanish Colonial Revival buildings from that era (poured slip-form concrete). They would stack 2-×-4s and pour the concrete in; then, with the 2-×-4s removed, one can see the grain of wood imprinted in the concrete surface, which was intentional.

One of the changes we have made with the new buildings is adapting a rather smooth finish, so this is one of the minor variations; but the roof tiles and the heavy dark wood that are featured are reminiscent of the James Complex. The changes we have made have been very intentional and carefully thought through, such as the added feature of the pavers at the Magnolia Avenue entrance onto Campus Bridge Drive. We tried to add a little interest, but at the same time remain respectful of the James Complex.

We have developed our details and color scheme from a core palette. We spent a lot of time and energy developing it. There is some flexibility in the design palate: for instance, we have approved different columns but they need to make sense. A couple of the wonderful things about this type of architecture are the arcades and breezeways. They are great features for a college, especially in a Mediterranean climate. One example of how we have used this concept can be found in the new Business Building, which has a breezeway that extends all the way through on the first floor. It even has large palm trees growing in it. It is quite impressive and gives you a marvelous indoor/outdoor feeling.

Moses It also looks as if you have shown a high level of respect for the mature landscape surrounding the original campus. How does that factor into your vision?

Ellis My wife and I love visiting Spain and Southern Europe. Spain especially has a similar climate to Riverside. In keeping with the climate and the Spanish type architecture, we are attempting to maintain a water-wise Spanish or Southwest style landscape, which draws on the mature plantings of the original campus, while going with a more xeriscape approach around our new buildings. It complements this type of architecture. Also, I love the cafe scene in Europe, with its emphasis on outdoor seating, so we have created lots of outside seating close to buildings, replicating the European cafe experience.

Two buildings coming online soon are the 153,600-square-foot Events Center and a 100,000-square-foot engineering building. We are following the same architectural and landscape theme for those buildings.

Moses Obviously, the Spanish Colonial Revival James Complex, by Jekel, and your trips with your wife to Southern Spain and Europe have had a great influence on what you want to accomplish architecturally and visually at CBU. Are there other historic buildings in downtown Riverside that have influenced your commitment to this architectural theme?

Ellis Certainly, number one is the Mission Inn. On our first visit, to interview for the job, the College put us up in a nice room at the Mission Inn. This was like "Oh wow! This is unbelievable." It was 1994, not long after the renovation. It was like a brand new old building. The Inn was fresh, and just a gem. Others that really stand out for me are the Municipal Auditorium, just a gorgeous building, a great sense of place; and Cafe Sevilla. It started as Johnson Tractor Company, designed by G. Stanley Wilson, and used the same slip-form concrete method as the James Complex by Jekel. It is just a beautiful little, simple, elegant and timeless building.

FIGURE 4 Eugene and Billie Yeager Center, CBU, completed in 2003. Broeske Architects & Associates, Inc.

Shortly after I arrived here, I was also introduced to an impressive Mission Revival complex west of Riverside that stuck in my mind: the Norconian Club. It blew me away. They gave us a tour, and although it was virtually closed, or about to be closed, the architecture made it a "wow" building with amazing space. The Norconian reinforced my experience at the Mission Inn. I love Las Campanas Mexican Restaurant, the Spanish Patio, and the Spanish Art Gallery; all unique spaces. Every time I turn a corner at the Mission Inn there is something marvelous that transports me to another time and place. This is what we have tried to do at CBU. Finally, I appreciate the core historic buildings in downtown Riverside. They represent the soul of the city.

Moses What is it about this type of architecture that elicits such a powerful response from you?

Ellis I think it is amazing what architecture can do to a person's mindset and emotions. It can put you in a wonderful mood. Spanish Colonial Revival architecture is romantic, in the sense that it is a fantasized ideal that creates an idyllic setting for enjoying life. Sitting outside among the flowers and greenery, drinking coffee, enjoying the company of others in the wonderful Mediterranean climate elevates the spirit. From a practical standpoint, it also creates a lot of visual interest.

Moses It's obvious that you have been quite systematic and thoughtful in the way you have approached the expansion of your campus, based on your strong views about the role of architecture to create a timeless and serene academic environment. How has your vision been received by parents, prospective students, and donors?

Ellis Comments from parents I receive at student orientations are usually very complimentary about the way the campus feels and how it meets their vision of a way a college should look and feel. We try to create an environment that feels like a timeless university. It is consistently telling you to calm down, slow down, and meditate on your studies and your spirit. I believe many parents, students, and donors appreciate the quality of what we have done here and see that the CBU campus suits the environment of Southern California. They feel it really fits here.

Moses How has the city of Riverside and the community received your vision?

Ellis We've received very good support from the city and even cheerleading. This was an area that was rather run down and we've tried to elevate it. We've made an effort to be sensitive to the surrounding neighborhood while taking what was here and enhancing it. Thus, the city has been very appreciative and has given us a thumbs-up to our new specific plan. It ties in with the Mission Inn, the heart and soul of what is traditional in Riverside, yet is modern. Our strategy seems to be working; it is bringing in thousands of students who have a huge economic impact, and they tend to be good citizens, too.

Moses In conclusion, Dr. Ellis, can you sum up for us what you see as the result of your efforts to emphasize the historic character of CBU's original campus, and your strategic plan of the University?

Ellis I have been here twenty-two years and it has just been a joy, and it continues to be my privilege to help shape an organization like CBU; especially since I deeply believe in its mission. From the first time my wife and I came to Riverside, we have been made to feel welcome. We were here only a couple of months before a reception was held for us in what used to be called the Music Room at the Mission Inn. There were so many people there welcoming us to the community. They gave me the opportunity to paint the vision of what we were going to do to turn around CBC. We were so well received, as was my plan. It was evident that Riverside is not just an exit on the freeway. There is a real history here and I like to tell visitors about it. It's our time now to honor that legacy, to be in the moment today, and to prepare another generation to take the baton. We hope they will appreciate the heritage at CBU and Riverside, and continue to value what a special place Riverside is and CBU's role in it.

Moses Thank you very much, Dr. Ellis, for speaking with us. Your love of CBU and its original Spanish Colonial architecture comes through loud and clear. We appreciate you sharing it with us.

Moses/Whitmore: Conclusion

By all appearances, Dr. Ron Ellis' turnaround plan for CBC/CBU, with its emphasis on the SCR architectural theme, and his application of modern theories of organizational change, has proven a success. He has taken a financially struggling liberal arts college of 808 students in 1994, to a fully accredited comprehensive university of 9,157 students in 2016. CBU currently contains eleven schools and colleges, including an online college.

In large measure, the CBU success story can be attributed to the impression Ellis received upon visiting the CBC campus for the first time. He immediately saw the value of Henry Jekel's SCR architecture, and its potential as a model for the permanence and timelessness he wanted to assign to the campus. A recent marker of his successful reading of the school, its architecture, and the community came in 2016, when CBU received its first $10 million gift to help fund a new 100,000-square-foot building to house the Gordon and Jill Bourns College of Engineering. That building, too, will embody Ellis' dedication to continuing the SCR in contemporary form.

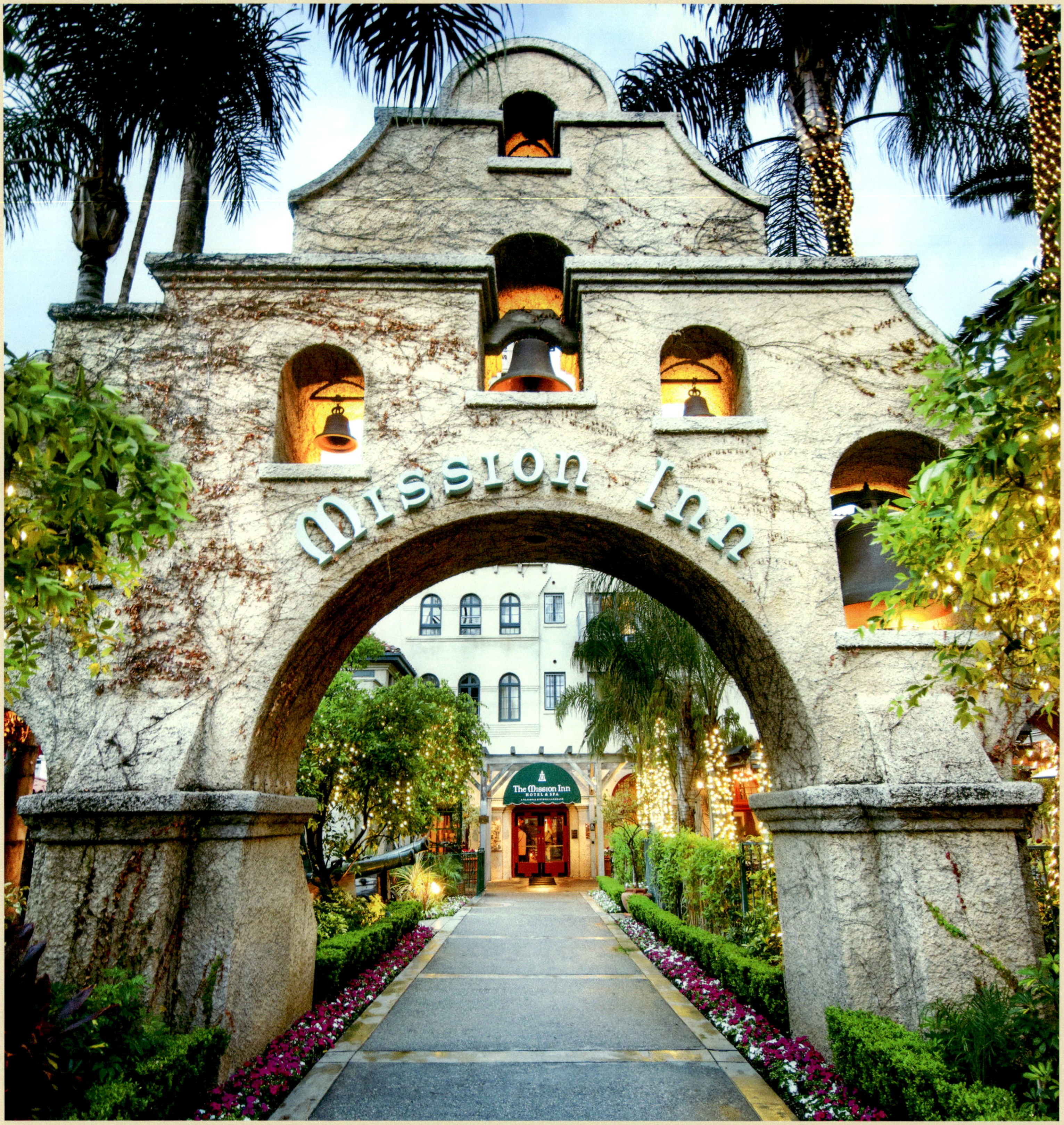
Mission Inn
The Mission Inn
HOTEL & SPA

7

Susan Straight

Bougainvillea, Lavandería, Cemetery, and Cross

Soldiers brought them from the ships, hung them first from trees, then on wooden frames. At last, the bells sounded from the campanario in the church itself—after we made it, after we built the church.
—Deborah A. Miranda, descendant of Chumash/Esselen peoples at the Santa Barbara, Santa Ynez and Carmel Missions, in *Bad Indians: A Tribal Memoir* (2013)

In 1970, my fourth grade class took a field trip to Mission San Gabriel. I was nine. The romantic history of the twenty-one California missions, with attendant lessons of structure, materials, and landscaping, as well as the ideas incorporated into Spanish Colonial Revival architecture, are deeply and curricularly ingrained in the state, into the very fingers and vision of children. For decades, all children who attend public elementary schools, and many who attend private schools, no matter their heritage, religion, citizenship, or native language, have been required during their fourth grade year to construct a California mission.

San Gabriel was the most beautiful and well-designed place I'd ever seen in my short life. Back in our garage, a stucco box with a corrugated metal slide-down door, separate from our 1960–built "modest ranch" tract home in Riverside, where five children shared two bedrooms, I was the first to build a mission. My dad, who loved design and construction, who later built two houses in the San Gorgonio Mountains, helped me. We measured and cut out from plywood the various buildings,

FIGURE 1 Model of Mission San Juan Bautista, constructed by Gaila Sims with her grandfather in 1998, Riverside

sawing archways into the walls, gluing them to a heavy plywood base. He taught me how to mix plaster of paris, and with a putty knife I spread the white onto the walls. As a nerd and perfectionist, I insisted on constructing a vineyard with twigs and dried tendrils of passion fruit vine that grew on our fence, and in the damp furrows of a nearby orange grove, even found moss that I attempted to transplant to real dirt we spread in the vineyard. I watered it all carefully. My dad and I carried the mission proudly into the classroom, where my obsessive personality was confirmed to teachers and fellow students even as the green moss turned to black velvet failure.

Twenty-eight years later, my dad and my neighbor assembled skilsaw and plaster of paris on my porch and helped my eldest daughter build Mission San Juan Bautista. I was a single mother of three daughters by then. It became a family tradition for each girl to choose two missions to visit, in anticipation of their project, and in April 1998, we carefully measured and sketched San Juan Bautista, laid out the garden plans and outbuildings. A facsimile of tile roof, plywood painted bright red, thick-smeared plaster, incongruous plastic doves bought from the wedding section, but an impressive courtyard and fountain enclosed by walls my daughter built of stone: gravel she chose painstakingly from our yard.

But two years later, for my middle daughter, though we would visit Mission San Carlos Borromeo de Carmelo, we first arrived in late afternoon at Mission La Purisima Concepcion, a more remote, almost austere mission outside of Lompoc. In the dark, chilly white-plastered infirmary, where we were told how many women died in childbirth, my girls shivered. In another outbuilding, with wrought-iron bars on the windows, we were told that native girls sometimes slept here in order to escape the Spanish soldiers. We walked outside the walls of the mission compound, where in the tall grass bending in the wind, we saw wooden crosses where native Californians had been buried, far from the consecrated cemetery where the bodies of the padres and Spanish rested. My middle daughter, at nine, said with a mix of anger and revelation, "This is where I'd be buried. Because I'm part Indian." She is Ethiopian, Swiss, French, Cherokee, and unnamed West African in heritage, as are her sisters. It was a moment none of us ever forgot, standing in the foxtails and wildflowers of Central California.

That spring, she built San Juan Capistrano, with tile roof made of lasagna sheets glued to cardboard and painted red, with painstaking detail of bell tower and façade, with more white doves. Years later, rats ate the roof, for which I'm very sad.

But that afternoon with my children, outside the walls of plaster blushing pink from setting sun, adobe bricks artfully visible in a worn corner, had changed the way I thought about missions, the history we're given, about our romantic obsessions with SCR architecture and design, and the icons which are meant to represent an Edenic past.

That summer, I went with my friend Gordon Johnson to the colonial outpost of Mission San Luis Rey de Francia. Gordon is a member of the Pala Band of Mission Indians: more than fifty distinct indigenous communities with their own languages, landscapes, belief systems and legends based in those landscapes, whether of seacoast or desert or chaparral-covered mountains or oak-shaded valley, more than fifty communities of peoples collectively called for centuries "Mission Indians" because of their proximity to those idealized compounds, named for people conscripted by force or desire or initial curiosity to come to these raw places of colonization, where zanja ditches were dug by hand and then lined with stone or tile, where massive oaks were cut for those lintels and roof beams.

Someone had to do the laundry.

Gordon and I walked around a deep sunken basin lined with hard-clay tiles and bricks, with openings where water would sluice down the narrow canals and into the stone troughs where countless women washed. Consider the laundry. Consider the filthy clothes of the humans building these places, then tilling fields for wheat and planting olive and fig and all the trees so beloved in our myths of the missions, and the grapevines for the fathers who made their own wine and used that red liquid to consecrate their meals and their "acolytes" as the California humans were called, in the communion of colonization. Consider the linens, the wine-dappled tablecloths, the bedding for hundreds, undergarments and napkins and blood-stained rags from battle and birth.

As we sat on a low stone wall looking into the sunken place where water would pool, Gordon said that his great-great-great-great grandmother had lived at San Luis Rey. "She could have been right here," he said softly.

Later he told me: "Her mission name was Maria Juana de Los Angeles, her married name was Maria Juana Sovenish, her maiden name was Maria Juana Yapeecha. Why she was called de Los Angeles I don't know. She was 4/4 Luiseno and maybe from the San Luis Rey Luisenos, because family stories place her at the mission in the early 1800s. She was the only Luiseno woman to be granted a rancho—Rancho Cuca." Gordon said, "There is a chair in my mother's house that she remembered sitting in at the ranch when she was a child."

Someone had to wash all those clothes, and even though the lavandería was marvelous in design, timeless and haunted as an ancient Roman ruin—San Luis Rey was built in 1798, one of the oldest missions, and places like this are California's Roman ruins—the images of women moving cloth over rough stones and

165

into the water were what I saw that day. I saw the infirmary, and the wooden crosses. I saw the women touching these same tiles not celebrated as iconic elements of SCR, telling each other stories and talking about who was in love, who was not in love, who was having a child, and who had lost one.

At the Mission Inn, in Riverside, Yuliana Buenrostros unloads hundreds of snowy white towels from her rolling cart each day along tiled corridors on the fourth floor, built in 1920s in the style of Granada, overlooking the Spanish Patio. "People think it's a mission, but then they find out it's a hotel," she said, laughing, taking a pile of linens into the next room. "Not a church."

The Mission Inn—not a religious outpost, but a boarding house and tavern that began as the Glenwood Springs Inn, the first wood and adobe bricks laid in 1880 by Christopher Columbus Miller (a fact, not irony), whose son is pictured as a child there. Frank Miller, that son, over later decades turned this place into a singular and eclectic architectural wonder. The first section of the Mission Revival-style building, which is now one of the most celebrated tourist and wedding destinations in Southern California, was designed by Arthur Benton between the years 1890 and 1901, before the explosion of SCR styles. But Frank Miller added more and more wings, in a fantastical mash-up of art and architectural styles from all over the world. Wandering the corridors inside, catacombs and passageways filled with hidden art and antique and whimsy, and the myriad walkways outside, visitors see Moorish domes from 1920 above the Granada Wing, and dine outside on the Spanish Patio below the red-tile overhang shading Yuliana, who is taking linens into Room 408.

FIGURE 2 Yuliana Buenrostros, Room 408, Mission Inn, Riverside

FIGURE 3 Mission Inn with clock tower, Riverside

FIGURE 4 Bell Tower in Spanish Court, Mission Inn, Riverside

She was born only a mile or so from here. Seventh Street was renamed Mission Inn Avenue some time ago; Buenrostros was raised on Ninth Street. She is twenty-six, the same age as my eldest daughter. They attended the same high school. Her father went to high school with me.

Her father came from Zacapa, Michoacan, Mexico, when he was sixteen; her mother came to Riverside when she was only five, from the same small village. They married, divorced, married other people, and twenty-two years later, remarried, which makes Yuliana laugh. She has ten siblings. She has never been to Michoacan, she says, smoothing her hands over the bedspread in the room with its careful SCR replica furniture and accents, her cart anchored near the wrought-iron balcony strung with crimson blossoms of bougainvillea. But she is going for the first time in a few weeks, to see her paternal grandparents, whose anchor in their tiny village is strong, with descendants all over Southern California.

For decades, the Arias Troubadors entertained patrons on the Spanish Patio below. Jose Arias, born in 1889 in Abasolo, Guanajuato, Mexico, came to Riverside when he was twenty-one, already playing guitar. He married Dolores Corral, who lived in the city, and with his four brothers-in-law, played every week at the Mission Inn, as well as entertaining at Mission San Juan Capistrano and eventually, when he was older, at the Ramona Pageant in Hemet, California. Arias was of the generation that knew original Spanish land grantees, the history of those ranchos, and their songs.

In the kitchen today, men and women of Mexican and Central American heritage prepare hundreds of crème brulees like tiny suns with burnt planetary crusts, murmuring in Spanish and English as they wield their torches

Coming up on the staff elevator, which has retained careful period detail, heading to the fourth floor corridor to tend those blossoms is Rafael, who was born near Universal Studios in North Hollywood to parents from Mexico. For seventeen years, he has maintained the

FIGURE 5 Jose Arias, Mission Inn, Riverside

FIGURE 6 Arias Family, Mission San Juan Capistrano, 1925

FIGURE 7 Delivering crème brulee, Mission Inn, Riverside

FIGURE 8 Rafael, supervisor, Mission Inn, Riverside

lush beauty of intensely specific romanticism, the bougainvillea carefully trained along pillar and pergola and balcony, white and red roses on patios and courtyards and rimming the fountains. Rafael came to Riverside because his wife's brothers were here. He is in charge of the third and fourth floors, the heart of the Granada Wing, at which thousands of diners gaze from the Spanish Patio. Behind him is the gleaming wood-faced clock that strikes bells on the quarter hour, which causes a rotating cast of male figures to appear in the arched opening: a Franciscan padre, then a Spanish soldier, then a Native Californian wearing an animal skin, pointing a spear at a bear.

The romantic remains in the fortunate and perfect melding of aesthetic—*Ramona*, the classic novel of early California, opens with the sounds of sparrows in the tangle of climbing blossoming vines trained over the wooden beams of a ramada. But readers forget that Helen Hunt Jackson intended readers to see the landscape not as Eden for the Native Californian characters persecuted, hunted, and killed for defiance and love, but for the Spanish who'd been granted their land in team with the shrewd and ruthless Americans.

It was always about land, before the beauty of design. Usually, invisible or briefly noted in curriculum of idyllic courtyards are the feet which blended earth and straw and hair into the adobe bricks for the walls that kept conscripted believers inside to work until the loveliness was finished, and kept free believers of other gods outside with their different sky.

Spanish—so often used in Southern California to draw a very clear distinction of architectural heritage and human lineage, setting the building's design and the human's ancestry as not Mexican, from that nation so close to here. Mexico shares artificial borders that have moved many times during the history of America, borders which remain quite porous despite walls, fences, and patrols. *Spanish* makes it clear that this design was conceived in Spain, the nation across the ocean and on a different continent.

Colonial—from colony. To colonize: to send a collection of humans away from their home to a place, usually far away, to establish an outpost at a distance, in order to amass land, to trade, to escape religious persecution or to bring religious persecution to others.

Revival—this can be the tent of evangelical religions, and the reanimation of something dead. Both, perhaps, in the fervor exhibited by architects who use SCR as a design accent applied lightly to countless brand-new and bland tracts of housing across California.

Alta California. Colonized eternally, murderously, artistically, and architecturally, and inextricably by many different nations. The perfect coalescing of architecture and style, climate and desire, and open space and money, the California people see even now in film and print as vision all over the world.

In 1774, Juan Bautista de Anza passed over the Santa Ana River a few miles from where Frank Miller later built the Mission Inn. Father Francisco Garces and Father Diaz, both Franciscan priests, joined Anza's party in San Xavier de Bac, a frontier mission in what is now southern Arizona. Anza, a Spanish military captain at Tubac, Arizona, wanted Spain to open a trading route with New California, a land route from Tubac to San Gabriel Mission. The procession must have looked like a strange parade to the Native Californians who saw them approaching, guided by a Native Californian who had previously made the trek from Arizona to San Gabriel, whose Christian name was Sebastian Tarabal. Twenty-one volunteer soldiers from Spain, an interpreter, a carpenter, five mule-drivers, two of Anza's servants. Sixty-five head of cattle, and 140 horses.

In 1775, Anza, Garces, and the Spanish organized a second crossing, this time partly to populate "New California." They brought 240 people, including 29 wives of the soldiers. They crossed deserts, lava flows like "a sea of broken glass," travelled down the Gila River to the

FIGURE 9 Rafael's bougainvillea on Spanish Patio, Mission Inn, Riverside

FIGURE 10 Padre in Anton revolving clock tower, Mission Inn, Riverside

land of the Yuma people, and finally crossed the Colorado River at a ford shown to them by a Yuma leader named Palma. The river ran two hundred yards wide and shallow enough that the cattle and horses were led across, while Anza and the others rode horseback. Father Garces, though, had a terrible fear of falling from a horse and drowning—he could not swim. So the Yumas carried the priest across the water and into California.

Father Junípero Serra became the most famous colonizer—legendary dispenser of faith and mustard seed along El Camino Real, named by those Spanish for their journeys to bring Catholicism to Alta California, he is immortalized in the garden of Mission Carmel in a bronze statue and canonized in 2015, his figure vandalized days after canonization with paint splashed by activists. Belief and fervor and horses and soldiers from Spain, coming up through Mexico, arrived in California to colonize people of different belief and fervor and bears and warriors who had lived in their ancestral lands since the stories of creation given to them by their elders.

There are kits now. For the most popular missions, the kits sell out early at craft stores in spring, when California schoolchildren are given their history assignment.

My youngest daughter had seen nine missions by her fourth grade year. She chose to visit San Miguel de Arcángel, in a tiny community off the old El Camino Real; when we arrived, the chapel and bell tower had been damaged by earthquake and were in disrepair, but schoolchildren were playing at the mission school, and women were preparing for Mass. Founded by Father Fermin Francisco de Lasuen in 1797 to close the gap between Mission San Luis Obispo to the south and Mission San Antonio to the north, by 1806 more than 1,000 Native Californians were living and working at San Miguel. The original buildings burned that

FIGURE 11 Mission kits

FIGURE 13 Figures to populate a mission

FIGURE 12 Model of Mission San Miguel de Arcangel, constructed by Rosette Sims in 2004, Riverside

FIGURE 14 Las Campanas bell tower, Mission Inn, Riverside

FIGURE 15 Roberto Loya in the bell tower at Our Lady of Guadalupe Shrine, Riverside

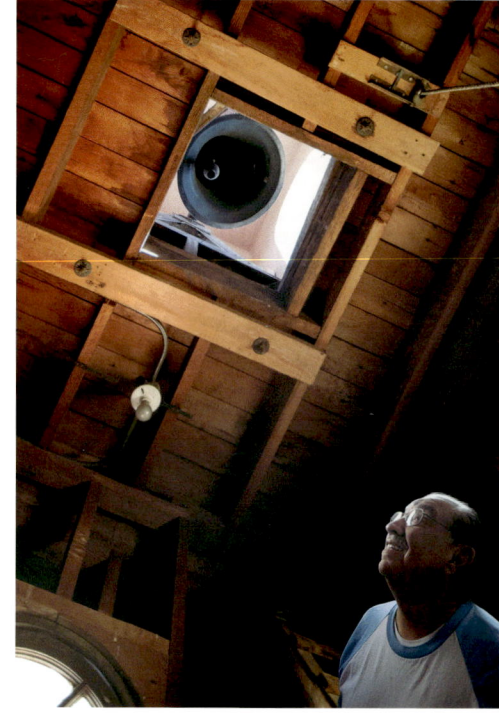

year—and a new adobe church was completed in 1821, with vivid and original interior frescoes and artwork we saw that afternoon. The place was emotional, the people we met at worship devoted to this place.

Independent and stubborn, my daughter made her San Miguel back home without much help. She mixed actual stucco for her walls, razored cardboard down to the corrugation for red-painted roof, and hung heavy bells of etched metal on her campanile.

This spring, kits from California Missions are displayed in a special island of their own at Michaels. What will nine-year-olds remember from red tile roofs, made of cardboard or lasagna or bought in a sheet from a kit? What do they feel as they position the "Padre with child," the gold-toned plastic cross, the "Indian male" who has an anachronistic tomahawk, the "Indian female" who has no laundry or linens at her feet, who may have been imprisoned in the infirmary for safety?

When we native-born Californians grow up and drive freeways along which we see the ubiquitous bell towers and patios of auto repair shops and fast food restaurants and plazas and malls and insurance agencies, do we believe this came from colonization or think of it as indigenous to California?

Belief is more complicated than that. The mission bells called people to work and pray and eat, no matter who they were or what they believed. The Catholic faith spread through the plains and coastal communities of California, and centuries later remains the genesis for this landscape of the every day. Referring to people or history as Spanish continues to differentiate them from Mexican or indigenous; immigration is now about different people crossing different rivers, not on horseback with priests, but sometimes with their shoes and clothing in plastic safekeeping on their heads while they wade or swim in order to wash the linens of hospitals, hotels, universities and factories, to

FIGURE 16 Ash Wednesday, Our Lady of Guadalupe Shrine, Riverside

cut the grass of said places, to repair the red tile roofs and plaster the weathered walls and solder the wrought-iron fences.

Some came last week. Some came a hundred years ago. Some come in airplanes or vehicles, and some on foot, and some were here already when Juan Bautista de Anza and Padre Junípero Serra arrived.

But millions in California, in America, live and worship and work and continue to build in a style that never truly belonged to anyone specific or identifiable, their hands from Mexico or El Salvador or Los Angeles or Santa Barbara or Riverside or Tijuana or Granada, still plastering and fitting red tiles into a pattern over binding cement, still pulling on an actual rope to ring an actual bell.

Las Campanas. Always inside the graceful domed bell towers, curves and arched openings and elegant flourish, sometimes three layers of bells to call the faithful and the exhausted at the missions, to proudly proclaim the entrance to the Mission Inn. Less than a mile from the hotel, Roberto Loya stands inside the bell tower at the Shrine of Our Lady of Guadalupe, the church built in the community where countless people of Mexican descent have lived in Riverside since the 1800s, where Yuliana Buenrostros grew up in sight of this church.

Loya's forehead is marked by a dark cross on Ash Wednesday as he looks up inside the scaffolding at the rope he has often pulled to ring the bell. "I probably spent 50 percent of my life here, at Our Lady," he said, gazing at the thick-flared edges of the bell above. Loya's father arrived in Riverside in the early 1900s, from Mexico, and the Loyas—including brothers Robert, Henry, and Prax—since their childhoods have raised money for this church.

A steady stream of people came through the arched gateway of Our Lady of Guadalupe, headed down the daylit center aisle, paused before the altar and received the cross of dark ashes on their foreheads. Babies whose skin had never been marked with soot, couples in their nineties who have been coming here since the church was constructed, and workmen recently arrived from Central America.

Our Lady of Guadalupe was built by people who made their own universe—out of wood-frame houses and new sidewalks, pepper trees and backyard lemons, trucks that came every morning to pick up workers for the citrus groves, while other workers headed downtown to the earlier incarnation of the Mission Inn and other hotels and restaurants.

In the 1890s, Saint Francis de Sales, in downtown Riverside, was the sole church for the entire parish area. Baptismal records show that 175 English-speaking families and 450 Spanish-speaking families attended. This city, as all cities in California, had strict residential boundaries then for Mexican-American and black residents. Families of color walked the long mile downtown in large groups for Mass, wearing their best clothes. By 1925 parishioners received permission from the diocese in Los Angeles to raise money for a "mission church." Built with local labor from future parishioners, although the building was not yet completed in 1929, the two bell towers still only wood-framed, the first Mass was held outside in the shade of the new walls.

Near those walls, Tony and Sarah Lopez sit for a moment. Sarah was a teenager during that first Mass in 1929. She was nineteen, a soloist in the choir, singing in Latin, when Tony Lopez, who'd come from Guanajuato, Mexico, when he was only two and lived nearby, became entranced. When he heard her voice that day, he said simply, "I felt something." They were married at the altar near where they now sit with shoulders touching. "Seventy-four years in August," Sarah said. Tony added, "I walked up here for thirty, forty years. I rang the bells to call the people to Mass. When someone died, I rang the bells to tell the people. A family asked me, 'Why do you ring the bell when someone dies?'" He gestured as if pulling hard on that rope, and then laughed. "I said, 'They should have come when I rang the bells on Sunday!'"

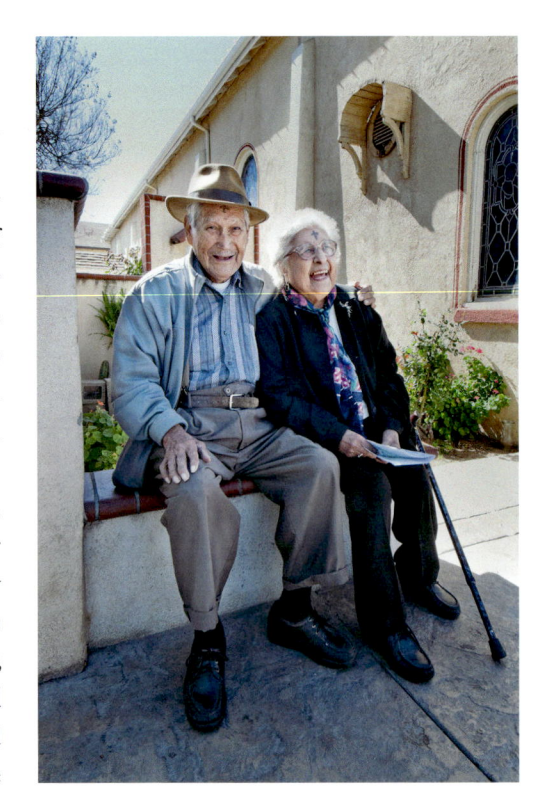

Downtown at the restaurant called Las Campanas, inside the Mission Inn where Frank Miller began his transformation of the boarding house in 1903 when he built the Mission Wing, employees shoulder trays bearing tortilla soup. Patrons glance up at Rafael's flowers from the Spanish Patio, designed along with the Alhambra Court by Myron Hunt in the 1920s as an ode to Granada. Brides pose for photos near the Carmel Wing, built from 1910 to 1915, where the ochre dome glows in Moorish style, which itself is Byzantium, originating in Turkey, and now in this Southern California skyline is meant to replicate the dome of Carmel Mission. Frank Miller wore the brown

FIGURE 17 Tony and Sarah Lopez, where they met and married and still worship, Our Lady of Guadalupe Shrine, Riverside

FIGURE 18   Moorish dome in Granada Wing, Mission Inn, Riverside

FIGURE 19 Frank Miller, owner of Mission Inn, dressed as a padre on Mount Rubidoux, Riverside, date unknown

FIGURE 20 Negotiating repairs to the tile roof, Mission Inn, Riverside

garments of a Franciscan father in photos from the nearby mountain which he purchased and where he erected a wooden cross. A workman hangs from the edge of one bell tower, replacing roof tiles after a hard rain, his location just above the brick wall built by Miller, a Quaker turned Congregationalist, who ordered pre-Columbian faces like gargoyles placed strategically along the boundary of his beloved inn. Yuliana moves her cart down the tiled corridor past the white roses to the next room on the fourth floor, and it is the one where I spent my own wedding night in 1983. I built a mission with my dad. My daughters and Yuliana built their missions so many years later, populated with the ghosts and the women carrying clean linens under the beautiful archways.

SELECTED READING LIST

Allen, Harris. "Spanish Atmosphere." *Pacific Coast Architect* 29 (May 1926): 5–7.

Almaguer, Tomas. *Racial Fault Lines: The Historical Origins of White Supremacy in California*. Berkeley: University of California Press, 1994. Reprinted with a new preface, 2009.

Architectural Resources Group. *Palm Springs Citywide Historic Resources Survey*. San Francisco, 2004.

Baxter, Sylvester, Bertram Grosvenor Goodhue, and Henry G. Peabody. *Spanish–Colonial Architecture in Mexico*. Boston: J. B. Millet, 1901. Reprint, Nabu Press (Public Domain), 2010.

Belloli, Jay et. al. *Myron Hunt (1868–1952): The Search for a Regional Architecture*. Los Angeles: Hennessey and Ingalls, 1984.

Belloli, Jay, Jan Furey Muntz et. al. *Johnson, Kaufman, Coate, Partners in the California Style*. Santa Barbara: Capri Press, 1992.

Bricker, Lauren Weiss, and Juergen Nogai. *The Mediterranean House in America*. New York: Abrams, 2008.

Brook, Vincent. *Land of Smoke and Mirrors: A Cultural History of Los Angeles*. New Brunswick, NJ: Rutgers University Press, 2013.

Cheek, Lawrence. "Taco Deco: Spanish Revival Revived." *Journal of the Southwest* 32, no. 4, (Winter 1990): 491–98.

DeLyser, Dydia. *Ramona Memories*. Minneapolis: University of Minnesota Press, 2005.

Deverell, William. *Whitewashed Adobe: The Rise of Los Angeles and the Remaking of its Mexican Past*. Berkeley: University of California Press, 2004.

Falassi, Alessandro, and Edward Tuttle. "California's Houses in Costume." *The Journal of American Folklore* 103, no. 410 (October–December 1990): 502–13.

Fu, A. S. "Materializing Spanish-Colonial Architecture: History and Cultural Production in Southern California." *Home Cultures* 92 (2012): 149–71.

Gebhard, David. "The Spanish Colonial Revival in Southern California, 1895–1930." *Journal of the Society of Architectural Historians* 26, no. 2 (May 1967): 131–47.

———. "Architectural Imagery, the Mission and California." *The Harvard Architecture Review* 1 (Spring 1980): 136–45.

———. "Tile, Stucco Walls and Arches: The Spanish Tradition in the Popular American House." In *Home Sweet Home: American Domestic Vernacular Architecture*. Edited by Charles W. Moore, Kathryn Smith, and Peter Becker (New York: Rizzoli, 1983).

———. "The Myth and Power of Place: Hispanic Revivalism in the American Southwest." In *Architectural Regionalism: Collected Writings on Place, Identity, Modernity, and Tradition*. Edited by Vincent B. Canizaro. New York: Princeton Architectural Press, 2007.

Gebhard, Patricia. *George Washington Smith: Architect of the Spanish Colonial Revival*. Layton, UT: Gibbs Smith Publisher, 2009.

Gelenter, Mark. *A History of American Architecture: Buildings in Their Cultural and Technological Context*. Hanover and London: University Press of New England, 1999.

Gellner, Arrol, and Douglas Keister. *Red Tile Style: America's Spanish Revival Architecture* (New York: Viking Studio/Penguin Group, 2002).

Gladding, McBean, Co. *Latin Tiles*. San Francisco: Taylor & Taylor, 1923.

Gonzales, Nathan. "Riverside, Tourism, and the Indian: Frank A. Miller and the Creation of

Sherman Institute." *Southern California Quarterly* 84, no. 3/4 (2002): 193–222.

Hall, Joan H. *Cottages, Colonials, and Community Places of Riverside, California*. Riverside, CA: Highgrove Press, 2003.

Herzog, Lawrence. *From Aztec to High Tech: Architecture and Landscape Across the Mexican–United States Border*. Baltimore: Johns Hopkins University Press, 1999.

Hines, Thomas S. *Architecture of the Sun: Los Angeles Modernism, 1900–1970* (New York: Rizzoli Press, 2010).

Howard, Carla Breer. "Casa Desert," *Desert Magazine* (March 2012): 88–92.

Hudson, Karen E. *Paul R. Williams, Architect: A Legacy of Style*. New York: Rizzoli, 1993.

Hunt, Myron. "First Congregational Church, Riverside, Cal." *The American Architect* 105, no. 2005 [May 27, 1914]: 116–22.

———. "Palos Verdes—Where Bad Architecture is Eliminated." *Pacific Coast Architect* 31 (April 1927): 9.

Klotz, Esther H., and Joan H. Hall. *Adobes, Bungalows, and Mansions of Riverside, California Revisited*. Riverside, CA: Highgrove Press, 2005.

Kropp, Phoebe S. *California Vieja: Culture and Memory in a Modern American Place*. Berkeley: University of California Press, 2006.

McCoy, Esther. *Five California Architects*. New York: Reinhold Publishing Corporation, 1960.

McGrew, Patrick. *Desert Spanish: The Early Architecture of Palm Springs*. Palm Springs: Palm Springs Preservation Foundation, 2012.

McMillan, Elizabeth Jean, and Melba Levick. *Casa California: Spanish Style Houses from Santa Barbara to San Clemente*. New York: Rizzoli, 1996.

Merchell, Anthony A., and Tracy Conrad. *Ojo Del Desierto: The Thomas O'Donnell House Palm Springs*. Palm Springs: The Willows Historic Palm Springs Inn, 2009.

Moses, Dr. H. Vincent, and Catherine Whitmore. *Henry L. A. Jekel: Riverside's Master Architect of Eastern Skyscrapers and the California Style, 1895–1950*. Riverside, CA: Inlandia Institute, 2017.

Neff, Wallace, Jr. *Wallace Neff: Architect of California's Golden Age*. Los Angeles: Hennessey and Ingalls, 1986.

Newcomb, Rexford. *Mediterranean Domestic Architecture in the United States*. Introduction by Marc Appleton. New York: Acanthus Press, 2000.

Ramón Lint Sagarena, Roberto. *Aztlán and Arcadia: Religion, Ethnicity, and the Creation of Place*. New York: New York University Press, 2014.

Rosa, Joseph. *Albert Frey, Architect*. New York: Rizzoli, 1989.

Starr, Kevin. *Inventing the Dream: California through the Progressive Era*. New York: Oxford University Press, 1985.

———. *Material Dreams: Southern California Through the 1920s*. New York: Oxford University Press, 1990.

Trafzer, Clifford E., Matthew Sakiestewa Gilbert, and Lorene Sisquoc, eds. *The Indian School on Magnolia Avenue: Voices and Images from Sherman Institute*. Corvallis: Oregon State University Press, 2012.

Weitze, Karen. *California's Mission Revival*. Los Angeles: Hennessy & Ingalls, Inc., 1984.

Whittlesey, Austin. *The Minor Ecclesiastical, Domestic and Garden Architecture of Southern Spain*. Preface by Bertram Grosvenor Goodhue. New York: Architectural Book Publishing, 1927. (Public Domain Reprint.)

Wilson, Chuck. *Quality Unsurpassed 1891–1991*. Glendora: California Portland Cement Company, 1991.

Winter, Robert, ed. *Toward a Simpler Way of Life: The Arts & Crafts Architects of California*. Berkeley: University of California Press, 1997.

Wyllie, Romy. *Bertram Goodhue: His Life and Residential Architecture*. New York: W. W. Norton, 2007.

BIOGRAPHIES

Aaron Betsky is president of the Frank Lloyd Wright School of Architecture at Taliesin, Wisconsin. Previously, he was director of the Cincinnati Art Museum and the Netherlands Architecture Institute. Betsky also curated the 11th Venice International Architecture Biennale in 2008. His most recent books are *Making It Modern* (2016) and *Architecture Matters* (2017). Mr. Betsky lived in California from 1985 to 2001.

Catherine Gudis is associate professor of history and director of the Public History Program at UC Riverside. She has worked for over twenty years with art and history museums, in historic preservation, and on place-based projects focused on California, including the co-curated exhibitions *Geographies of Detention: From Guantánamo to the Golden Gulag* at the California Museum of Photography, and *Junípero Serra and the Legacies of the California Missions* at the Huntington Library. The author of *Buyways: Billboards, Automobiles, and the American Cultural Landscape* and editor of books addressing visual culture and the urban landscape, Gudis is also co-founder of the art collective Project 51, which invites audiences to reimagine and reclaim the 51-mile LA River. Currently she is piloting a project on migrations and immigration as part of a partnership with the California State Parks. Gudis has held fellowships at the Getty Research Institute, Harvard University, and the Smithsonian Institution, and is at work on a book entitled *Curating the City: The Framing of Los Angeles*.

Douglas McCulloh is a photographer, writer, and curator. He is a four-time recipient of support from the California Council for the Humanities and has curated fifteen exhibitions, including three for the California Museum of Photography. His own exhibition record includes Victoria and Albert Museum, London; Central Academy of Fine Arts, Beijing; Musée de l'Elysee, Lausanne; Musée Nicéphore Niépce, France; La Triennale di Milano, Italy; Centro de la Imagen, Mexico City; Art Center College of Design, Los Angeles; Smithsonian Institution, Washington DC; and Cooper Union School of Art, New York. McCulloh's fifth book, *The Great Picture: Making the World's Largest Photograph* was published in 2012 by Hudson Hills Press. Exhibitions curated by McCulloh have shown in a range of venues: Kennedy Center for the Arts, Washington DC; Canadian Museum for Human Rights, Winnipeg; Centro de la Imagen, Mexico City; Flacon Art Center, Moscow; Center for Visual Art, Denver, Colorado; Manuel Álvarez Bravo Center, Oaxaca; Sejong Center, Seoul, South Korea; Central Academy of Fine Arts, Beijing, China; and Peterson Automotive Museum, Los Angeles.

Patricia A. Morton is associate professor in the art history department at University of California, Riverside, and is the author of *Hybrid Modernities: Architecture and Representation at the 1931 International Colonial Exposition in Paris* (MIT Press, 2000). She has lectured and published widely on modern and contemporary architectural history and race, gender, and identity. Her current research examines public space and postmodern architecture, focusing on the work of Charles Moore. Morton is editor of the *Journal of the Society of Architectural Historians*.

H. Vincent Moses, PhD, lives in Riverside with wife Catherine Whitmore, where they own and operate Vincate & Associates Historical Consultants, specializing in historic preservation and museums. Retired director of the Riverside Metropolitan

Museum, Moses focuses on the diverse cultural history and architecture of the region. At RMM, he supervised the transfer of the National Historic Landmark Harada House to the City of Riverside, and served as consulting museum historian for the planning and development of California Citrus State Historic Park in Riverside. With his wife Cate, Moses is co-author of *The Victoria Club: A Centennial Edition* (2003). Their most recent book, *Henry L. A. Jekel: Master Architect of Eastern Skyscrapers and the California Style, 1895–1950* was just released in October 2017.

Lindsey Rossi, Getty Foundation Curator of *Myth & Mirage*, holds an MA in decorative arts, design history, and material culture from the Bard Graduate Center in New York, and a BA in art history from UCLA. She is an independent curator who specializes in the history of design and decorative arts and focuses on nineteenth- and twentieth-century revivalist styles. Rossi has curated several design and architecture exhibitions for the Riverside Art Museum. She has contributed extensive curatorial research for the Metropolitan Museum of Art and assisted on design exhibitions and publications at the Los Angeles County Museum of Art and the Bard Graduate Center Gallery.

Carolyn Schutten is a visiting fellow for the Center for US–Mexican Studies at UC San Diego and is a PhD candidate in history at UC Riverside. She holds an MA history in museum curatorship from UC Riverside and an MA in urban and regional planning from California State Polytechnic University, Pomona. Schutten has received awards from UC Institute for Mexico and the United States, UC California Studies Consortium, UC Humanities Research Institute, UCLA Institute for Research on Labor and Employment, California Planning Foundation, and Blum Initiative on Global and Regional Poverty. During her tenure at the Riverside Art Museum, Schutten curated several exhibitions, including *Self Help Graphics: Aztlán, the Permanent Collection, and Beyond*. Schutten sits on the Public Art Advisory Committee at the Arts Council for Long Beach and has served as Inland Empire Regional Councilmember for the California Association of Museums.

Susan Straight has published eight novels, including *Between Heaven and Here* (McSweeneys, 2013), the final book in the Rio Seco trilogy. *Take One Candle Light a Room* (Anchor Books) was named one of the best books of 2010 by *The Washington Post* and the *Los Angeles Times*, and *A Million Nightingales* (Anchor Books) was a Finalist for the *Los Angeles Times* Book Prize in 2006. *Highwire Moon* was a Finalist for the 2001 National Book Award. "The Golden Gopher," published in *Los Angeles Noir*, won the 2008 Edgar Award for Best Mystery Story. Her stories and essays have appeared in *O. Henry Prize Stories, Best American Short Stories, Best American Essays, The New York Times, The New Yorker, The Los Angeles Times, Harper's, McSweeneys, The Believer, Salon*, and elsewhere. Her awards include The Kirsch Award for Lifetime Achievement from the *Los Angeles Times*, The Lannan Prize for Fiction, a Guggenheim Fellowship, and the Gold Medal for Fiction from the Commonwealth Club of California. She is Distinguished Professor of Creative Writing at UC Riverside. Straight was born in Riverside, where she lives with her family.

Catherine Whitmore lives in Riverside with her husband H. Vincent Moses. They have three sons and two grandsons. Cate worked as curator of anthropology for the Riverside Municipal Museum, and then as curator of history for the Riverside County Regional Parks and Open Space District. Retired from the county, Cate is co-owner with her husband of Vincate & Associates, a consulting business specializing in historic preservation and museums. Their most recent book, *Henry L. A. Jekel: Master Architect of Eastern Skyscrapers and the California Style, 1895–1950* was just released in October 2017.

INDEX

Note: Page numbers in italic type indicate illustrations.

adobe, 115–16, *119*, 124, 145
aggregates, 112–13
A. K. Smiley Public Library, Redlands (Griffith), 50–51, *51*, *52*, 117
Alberhill Coal and Clay, 116–17, 124, 127
Albert Burrage Mansion (Monte Vista), Redlands (Brigham and Coveney), 51, 53, *53*, *114*, 117
Alberto's/Circle K, Pomona, *146*
Allen, Harris, 119
American Architect (magazine), 56
American Institute of Architects, 54
Andalusian Farmhouse Vernacular, 63–68
Anderson, E. A., 66
Anglo-American peoples and ethos, 16, 19–20, 27, 30–31, 34, 37, 43–45, 56, 59, 84, 112, 116, 119
Anton Clock, 82–83, *83*, *170*
Anza, Juan Bautista de, 83, *83*, 170–71
Archer, John, 142
arches, 145–46, *147*
architectural style, and race, 11–14
Arias, Jose, and family, 168, *168*
Arts & Architecture (magazine), 31, 139
Arts and Crafts movement, 34, 44, 46, 48, 50, 57, 64, 117
Aubury, Lewis, 116
Automobile Club of Southern California, 88

Balch & Stanbery, Fox West Coast Theatre, Riverside, 58–59, *58*
Bandini, Juan, 115, 119
Banham, Reyner, 27, 140
Batchelder, Ernest, 117
Beattie, George, 124
Beaux-Arts style, 32, 33, 35, 54, 55, 58, 64, 65
Bell, Glen, 143
Benedict Castle (Castillo Isabella), Riverside (Jekel), 62–63, *63*
Benedict, Charles W., 62–63
Benton, Arthur, 46, 54, 102; First Church of Christ, Scientist, Riverside, 48, *49*, *139*; Memorial and Municipal Auditorium, Riverside, 56–57, *57*; Mission Inn, Riverside, 20, 32, 45, *45*, 47–48, *47*, 58, 78, 80, *81–84*, 82–86, *86–89*, 88, 90–91, 101–9, *102–5*, 145, 158–59, *162*, 166, *166*, *167*, 168–69, *169*, *170*, *172*, 174, *175*, *176*; Raincross symbol, 90
Beverly Hills Civic Center (Moore Ruble Yudell), 36
Boller, Carl, Corona Theatre, Corona, 61, *61*
Bolton, Herbert Eugene, 113
Bonheur, Rosa, 85
Borton, Francis S., 85
Boyd, William and Laura, 50
brick, 116–18, 124
Brigham, Charles, Monte Vista (Burrage Mansion), Redlands, 51, 53, *53*, *114*, 117
Broeske Architects & Associates, Eugene and Billie Yeager Center, California Baptist University, *159*
Buenrostros, Yuliana, 22, *166*, 168, 176
bungalows, 30, 32, 134, *136*
Burnham, Franklin Pierce, Carnegie Library, Riverside, 48, 50
Burrage Mansion. *See* Albert Burrage Mansion (Monte Vista), Redlands (Brigham and Coveney)
Byers, John, 32
Byzantine style, 48

California Baptist University, 20–21, 153–61, *154*, *156*, *159*
California Building (Goodhue), 56
California Fruit Growers Exchange (Sunkist), 45
California Portland Cement Company, *112*, *117*, 119, 121–22, 125, 127
California Studies, 19
California Style, 44, 50, 54, 58, 62, 66, 68
California Theatre, San Bernardino (Perrine), 60–61, *60*
Californios, 16, 43, 78, 114
Cano, Atilano, 125
Carnegie Library, Riverside (Burnham), 48, *50*
Carrere and Hastings, Hotel Alcazar, St. Augustine, 32
Casa Blanca School, Riverside (Wilson), *122*, 126
Casa de Anza Hotel, Riverside (Wilson), 65–66, *65*
Casa Palmeras Apartments, Palm Springs (Williams), 68
Case Study House Program, *Arts & Architecture*, 31, 139
Cassidy, Hopalong (William Boyd), 50
Catholic Church, 19, 29, 44, 88
Cecil Brashears house, Redlands, 66
cement, 118–25, 127, 145
Cheek, Lawrence, 142–44
Cheney, Charles, 56
Chicanos, 14
Chumash, 18
Churrigueresque style, 55–56, 80, 88, 92, 105
City Hall, Riverside (Jones), 58
City of Riverside Garage Seven, Riverside, *147*
civic buildings, 34–36
Civic Center, Riverside, 56–58, 102
Classical style, 48
Clements, Stiles O., 127
colonialism, 17, 170
Colonial Revival, 32, 33, 34
Community Settlement House, Riverside, 126
concrete, 112–13, 118–19, 124, 158
Corona, 44, 61
Corona High School, Corona (Wilson), 61, *61*
Corona Theatre, Corona (Boller), 61, *61*
Corydon Construction, 127
courtyard houses, 61–62
courtyards, 17–18, *18*, 30, 35, 47
Coveney, Charles C., Monte Vista (Burrage Mansion), Redlands, 51, 53, *53*, *114*, 117
C. P. Hancock & Son, 118, *120*, 127
Craftsman style, 33, 134
Cram, Ralph Adams, 31
Cresmer Manufacturing Company, 50, 61, *100*
CVS Pharmacy, Riverside, *144*, *147*

181

DaVita Pomona Dialysis Center, Pomona, *147*
Della Robbia bas-reliefs, 86, *86*
Desert Spanish Colonial style, 68
Deverell, William, 115–16, 124, 127
domes, 145, *147*
Donut Avenue, Chino, *14*
Double Indemnity (film), 33

École des Beaux-Arts, Paris, 54
El Camino Real commemorative bells, *10*, 88, 90
Elijah Parker Residence, Riverside (Spurgeon), 65, *65*
Ellis, Ron, 20, 153–61, *154*
El Mirador Hotel, Palm Springs, *12*
Escondido Art Center (Moore Ruble Yudell), 36
espadaña gable, 146, *147*

Falassi, Alessandro, 144
Fandango scene, *15*
Farrell, Charles, 60
Featherstone, Marion, 67
Federal Post Office, Redlands (Wilson), 59–60, *59*
Federal Post Office, Riverside (Taylor), 58, *115*
film, 33
First Christian Church, Riverside (Wilson), 68, *69*
First Church of Christ, Scientist, Riverside (Benton), 48, *49*, *139*
First Congregational Church, Riverside (Hunt), 54–56, *55*, *56*, 58, 78, 92–93, *92*, *93*, 119
Forty, Adrian, 142
Fox West Coast Theatre, Riverside (Balch & Stanbery), 58–59, *58*
Franciscans, 20, 29
Francis of Assisi, Saint, *83*
French Provincial style, 33
Fu, Albert, 119, 137, 145
Funk, Tom, "Everyday Tastes From High-Brow to Low Brow," *136*

Garces, Francisco, 170–71
Garcia, Mario T., 113
Gaynor, Janet, 60
Gebhard, David, 32, 46, 54, 61, 138, 139, 144, 145
Gebhard, Patricia, 64
Gehry, Frank, 31
Gill, Irving, 124
Glenwood Cottages, Riverside, 20, 80, 101, *102*. See also Mission Inn, Riverside (Benton)
Glenwood Hotel, Riverside, 45, 80. See also Mission Inn, Riverside (Benton)
Glenwood Tavern, Riverside, 101. See also Mission Inn, Riverside (Benton)
Goff, E. F., 80, 82

Gonzales, Nathan, 124
Goodhue, Bertram, 35, 54, 56, 124; California Building, 56
Graves, Michael, library, San Juan Capistrano, 30
El Greco, 85
Greene Brothers, 30
Greenough, Mary, 63
Grey, Elmer, 119
Griffith, T. R., A. K. Smiley Public Library, Redlands, 50–51, *51*, *52*, 117
Guerin, Jules, 35

Hammond House, Riverside (Jekel), 62
Hearst, Phoebe Apperson, 91
Hearst, William Randolph, 85
Heisler, Howard G., 64
Hickok, Clinton, 48, 50
Hogue, Bill, 154
The Home Depot, Pomona, *135*
Hosp, Franz P., 53
Hotel Alcazar, St. Augustine (Carrere and Hastings), 32
Huizing, Garret, 67
Hunt and Grey, 54–55, 119
Hunt, Myron, 54–55; First Congregational Church, Riverside, 54–56, *55*, *56*, 58, 78, 92–93, *92*, *93*, 119; Spanish Wing, Mission Inn, Riverside, 82, 102, 174
Hunt, Sumner, 46
Huntington, Henry E., 20, 46, 102

Indian boarding schools, 20, 83, 122–23
Indians. *See* Native Americans
Industrial Workers of the World, 125
Inland Empire, 19–22, 43–45, 47, 68, 77–78, 93, 95, 112–13, 116–20, 126–27, 133, 142, 145, 154
Italian Renaissance Revival, 91

Jack in the Box, Pomona, *146*
Jackson, Helen Hunt, 19; *Ramona*, 33, 43, 44, 47, 77, 78–79, 122, 124, 169
Jackson, J. B., 30
Jekel, Henry L. A., 20, 33, 56, 62–64, 69, 119, 120; Benedict Castle (Castillo Isabella), Riverside, 62–63, *63*; California Baptist University, 153–54, *154*, *156*, 157, 161; Hammond House, Riverside, 62; Krinard Residence, Riverside, 62, *62*
Jensen, Cornelius Boy, 115
Johnson, Gordon, 22, 165
Jones, Howard, City Hall, Riverside, 58
Judson, Horace, Sistine Madonna window, 92, *92*

Kaufman, John, 32
Keith, William, 85
Kelso Depot & Clubhouse, 78, 93, 95
Krinard Residence, Riverside (Jekel), 62, *62*
Kropp, Phoebe, 111–12, 133, 138, 145, 148
Kumeyaay, 18

labor, 21, *110*, 112–27, *112–15*, *117–21*
Lamar, Welmer P., 50
Landmarks Club, 19, 45, 46
Land of Sunshine (journal), 19, 45
La Plaza, Palm Springs, *14*
Lasuen, Fermin Francisco de, 171
Leslie Harris house, Redlands, 66
Limbert, Charles, 86
Lippich, Richard, baptismal font, 92, *93*
Lipsitz, George, 17
Lopez, Tony and Sarah, *174*, *174*
Loring Theater and Office Building, Riverside (Wilson), 58
Los Angeles Pressed Brick Company, 117
Los Angeles Times (newspaper), 43, 45, 46
Loya, Roberto, 22, *172*, *173*
Lummis, Charles Fletcher, 19, 43, 45–48, 51, 101, 115–16
Lynes, Russell, 139, 140; "Highbrow, Lowbrow, Middlebrow," 134, *136*, 137

Maria Juana de Los Angeles, 22, 165
Markham, H. H., 116
The Mark of Zorro (film), 33
Maybeck, Bernard, Palace of Fine Arts, San Francisco, 35
May, Cliff, 139; Mondavi Winery gateway, 36
McCulloch, Hugh, 66–67
McCulloch, Robert, 66–67
McCulloh, Douglas, 148
McGroarty, John, 19
McMillian, Elizabeth, 148
McNeeley, W. H., 121
McWilliams, Carey, 14, 16, 19
Mediterranean Revival, 54–62, 64–66, 102–3
Memorial and Municipal Auditorium, Riverside (Benton and Wilson), 56–57, *57*
Mendosa, Jesus, 125
Mexican peoples: actual and symbolic dispossession of, 13–14, 17, 21, 43, 111–14; as laborers on SCR architecture, *110*, 112–14, *112–14*, 117, 119, 120–27; living conditions of laborers, 124–26; in Southern California, 115; Spanish peoples distinguished from, 16, 115, 138, 169, 172
middle class, 19, 28, 31, 32, 33, 35, 36, 37
Miller, Christopher Columbus (C. C.), 80, 101, 166
Miller, Frank, 20, 43–48, *45*, 51, 54, 56–57, 79–80, 82–86, 88, 90–91, 95, 101–9, 123, 166, 174, 176, *176*
Miller, Isabella Hardenberg, *45*
mineral resources, 116
Miranda, Deborah A., 163
Mission Bridge Brand orange box label, *44*
Mission Carmel, 171
mission churches, 29
Mission Cult, 43, 47–48, 51, 101

Mission furniture, 46, 86, 93, 134
Mission Indian Federation, 51
Mission Indians, 165
Mission Inn, Riverside (Benton), 20, 32, 45, *45*, 47–48, *47*, 58, 78, 80, *81–84*, *82–86*, *86–89*, 88, 90–91, 101–9, *102–5*, 145, 158–59, *162*, 166, *166*, *167*, 168–69, *169*, *170*, *172*, 174, *175*, *176*
mission kits, 171, *171*
Mission La Purísima Concepción, 18, 21, 164–65
Mission Revival, 45–53, 91, 93, 102, 117
missions, 29–31, 43, 45, 79, 88
Mission San Antonio de Pala, 80, 82
Mission San Gabriel, 18, 48, 80, 124, 163
Mission San Juan Bautista, *164*
Mission San Juan Capistrano, *18*, 80, 165
Mission San Luis Rey de Francia, 21, 80, 165–66
Mission San Miguel de Arcángel, 171–72, *171*
Mission Santa Barbara, 18, 57, 58, 93
Mission Santa Inés, 18
Mission Village, Riverside, *135*
Mizner, Addison, 33
Modernism, 68, 138–39
Mohr, W. A., Santa Fe Depot, San Bernardino, *13*, 53, *53*, 78, 93, *94*
Mondavi Winery gateway (May), 36
Monsanto House of the Future, Disneyland, 138–39
Monte Vista (Albert Burrage Mansion), Redlands (Brigham and Coveney), 51, 53, *53*
Montgomery, Margaret, Great Cross, 93, *93*
Moore, Charles, 133, 140; Anawalt house, Point Dume, 142
Moore, Lester S., 54
Moore Ruble Yudell: Beverly Hills Civic Center, 36; Escondido Art Center, 36
Moorish style, 46, 48, 50–51, 57, 60
Mooser, William, III, Santa Barbara County Courthouse, 29, 30, 35–36
Morgan, Julia, 85, 91; YWCA (now Riverside Art Museum), Riverside, 57, *58*, 78, 91–92, *92*, 118
Morris, William, 134
Morton, Patricia, 69
Moss, Henry, 142
Mr. Blandings Builds His Dream House (film), 32
Mumford, Lewis, 134
Murillo, Bartolomé Esteban, 85
Murphy, Paul Edgar, 145

National Register of Historic Places, 44, 57–61
Native Americans: actual and symbolic dispossession of, 13–14, 16, 17–18, 21, 43, 78, 112; boarding schools for, 20, 83, 122–23; as laborers on SCR architecture,

122–24; Mission Indians, 165; resistance and insurgency of, 17, 18
Neff, Wallace, 32
Neighbors of Woodcraft, 20
neo-SCR, 21
Neutra, Richard, 31, 68, 137
Newton, Ryland A., 67
North, John Wesley, 47

Office Max, Pomona, *148*
Old Adobe of Mission Inn, Riverside, *15*
Otis, Harrison Grey, 46–47
Our Lady of Guadalupe Shrine, Riverside, *118*, *123*, 126, *172*, *173*, *174*, *174*
Ovnick, Merry, 139

Pala Band of Mission Indians, 165
Palace of Fine Arts, San Francisco (Maybeck), 35
Palm School, Riverside (Wilson), *61*
Palm Springs Preservation Foundation, 68
Panama-California Exposition (San Diego, 1915), 35, 45, 56, 78, 85, 124
Panama-Pacific Exposition (San Francisco, 1915), 35
Pasadena City Hall, 35
Perrine, John Paxton, California Theatre, San Bernardino, 60–61, *60*
Peyri, Antonio, 80
Pius X, Pope, 86
Playa Vista, Los Angeles, 37
Pomona Marketplace, Pomona, *147*
Postmodernism, 21, 133, 140
Pratt, Richard, 122
Price, William, 118–19
promotional materials: California Parlor Car Tours Company brochure, *15*; Casa de Anza Hotel, Riverside, advertisement, *65*; Los Angeles County pamphlet, *16*
Protestantism. *See* Anglo-American peoples and ethos
Pueblo Revival, 47

Rabbeth, W. E., Robert McCulloch residence, Redlands, 66–67, *67*
race: architectural style and, 11–14; building materials and, 115; SCR and, 12, 17, 111–12, 116, 138
Raincross Promenade, Riverside, *147*
Raincross symbol, 90, *90*
Ramona Outdoor Play, 78, *79*
Ranch style house, 36, 134, 138, 139, *140*
Redlands, 44, 50–53, 59–60, 66, *66*
Redlands Bowl, 59, *59*
refried architecture, 21, 143–44
Remington Estates, Chino, *14*
residential design, 61–68
Ribera, Jusepe de, 85
Richardsonian Romanesque, 46
Ridgecourt (Hickok residence), Riverside (Wilson), 48, 50, *50*

Rite Aid, Riverside, *141*
Riverside, 20, 33, 44–50, 54–59, 77–78, 90–92, 101, 159–60
Riverside-Arlington Heights Fruit Exchange, Riverside (Wilson), 58
Riverside Art Museum (Morgan), 57, *58*, 91–92, *92*, 118
Riverside Portland Cement Company, 116, *116*, 120, 122, 124, 125, 127
Robert McCulloch residence, Redlands (Rabbeth), 66–67, *67*
Robidoux, Flora, 124
Robidoux, Louis, 115, 124
Roosevelt, Theodore, 45, 48, 102

Sagarena, Roberto Ramón Lint, 16, 111
San Bernardino, 44, 53, 60–61, 66, *66*
San Bernardino Asistencia, Redlands, *10*, *119*, 123–24
San Bernardino County Historical Society, 124
Sánchez Coello, Alonso, 85
Sandham, Henry, 32
San Diego, 54, 56
San Juan Capistrano library (Graves), 30
Santa Ana River bridge, *44*
Santa Barbara, 44
Santa Barbara County Courthouse (Mooser), *29*, 30, 35–36
Santa Fe Depot, San Bernardino (Mohr), *13*, 53, *53*, 78, 93, *94*
Santa Fe Railroad, 47
Schindler, Rudolf, 30–31, 68, 137; Schindler-Case house, West Hollywood, 30
Schupach, Fred, 119
SCR. *See* Spanish Colonial Revival
secularization, 19, 44
Serra, Junípero, 19, 29, 80, 82, 90, 171
Sheets, Millard, Great Cross, 92, *93*
Shell, Glen Avon, *135*
Shepley Bulfinch Rutan, Stanford University, 32
Sherman Institute, United States Indian School, Riverside, 20, *83*, 122–23
Shingle Style, 32
Simons Brick Company, 124
Sims, Gaila, model of Mission San Juan Bautista, *164*
Sims, Rosette, model of Mission San Miguel de Arcángel, *171*
Slover, Isaac, 115, 116
Slover Mountain, 116, 119
Smiley, Albert K., 51
Smiley, Alfred, 51
Smith, George Washington, 32, 63–64
Southwest Museum, 45
Spanish Colonial Revival (SCR): elements of, 21, 29–32, 44, 144–46; as fantasy, 13–14, 16, 17, 19–20, 28, 43, 46, 83–84, 95, 101, 133, 138, 144–45, 148, 160; meanings of, 16, 18–19, 20, 27–28, 33–34; Mediterranean

Revival phase, 54–62; Mission Revival phase, 45–53; neo-SCR, 21; popularity of, 33, 44, 47, 69; postwar, 138–49; race and, 12, 17, 111–12, 116, 138; significance of, 12; sources for, 29; variations in, 28; as vernacular style, 36–37, 137; violence represented by, 17–18
Spanish peoples and ethos, 16, 115, 138, 169, 172
Spanish Renaissance Revival, 54–55, 105
Spurgeon, Robert, Jr., 33, 65, 69; Casa de Arroyo, Riverside, 65; Elijah Parker Residence, Riverside, 65, *65*; Oaklodge, Riverside, 65; residence, Riverside, *26*; Roberts Leinau residence, Riverside, 65
Stalder Building façade, Riverside (Wilson), 58
Stanford University (Shepley Bulfinch Rutan), 32
St. Anthony's Church, Riverside, *121*, 126
Starbucks, Riverside, *132*
Starr, Kevin, 20, 43, 48, 101
St. Francis of Assisi Chapel, Mission Inn, Riverside, 88, *89*, 106
Stickley, Gustav, 46, 86, 134
Stick Style, 30, 32, 48
stucco, 145
Sullivan, Louis, 46
Sunset (magazine), 33, 139
Sunset Tile Company, 118

Taco Bell, 133, 143–44, *143*, 146
Tanner, William Charles, 85
Tarabal, Sebastián, 170
taste, 134, *136*, 137
Taylor, James K., Federal Post Office, Riverside, 58, *115*
Taylor Brothers Brick Company, 116, 117, 127
Tibbets, Eliza, 48
Tiffany, Louis Comfort, rose window, 88, *89*
tiles, 34, 80, 82, 118, 145
tilt-slab construction, 124
Torres-Rouff, David, 126
Toys R Us, Pomona, *147*
Trabajadores Unidos (union), 125
train depots, 93, 95
Trujillo, Lorenzo, 115
Tucker, George E., 121–22
Tuck, Russell, 154
Tudor style, 33, 34
Tuttle, Edward, 144

UEI College, Riverside, *138*
Union Pacific Bridge, Riverside, *117*, 119–20
Union Pacific Depot, Riverside, *46*, 47
Union Pacific Railroad, 47
University Village, Riverside, *138*, 142
U.S. Department of Agriculture, Gardens and Grounds Division, 50

Vasquez, Preciliano, 121
Venturi, Robert, 144
Village Liquor, Chino, *146*
Villanuevo, Lucy, 124–25
Vitruvius, 53, 54

Walgreens, Riverside, *12*
walls, 27–30
Warner, Charles Dudley, 54
Wayne, Allen, 154
Weber, Peter, 103
White, Florence and Clarence, 59
Whiteson, Leon, 137, 142
Der Wienerschnitzel, Chino, *135*
Williams, Paul R., Casa Palmeras Apartments, Palm Springs, 68
Willie Boy, *113*
Wilson, G. Stanley, *100*, 102–3, 120; Atrio of Saint Francis, Mission Inn, Riverside, 105, *106*, 109; Author's Row, Mission Inn, Riverside, *104*; Cafe Sevilla (Johnson Tractor Company building), 158–59; Casa Blanca School, Riverside, *122*, 126; Casa de Anza Hotel, Riverside, 65–66, *65*; Corona High School, Corona, 61, *61*; Federal Post Office, Redlands, 59–60, *59*; First Christian Church, Riverside, 68, *69*; International Wing, Mission Inn, Riverside, 59, 64–65, *100*, 102–3, 105, *107*; Loring Theater and Office Building, Riverside, 58; Memorial and Municipal Auditorium, Riverside, 56–57, *57*; Palm School, Riverside, *61*; proposed additions, Mission Inn, Riverside, *105*; Ridgecourt, Riverside, 48, 50, *50*; Riverside-Arlington Heights Fruit Exchange, Riverside, 58; Stalder Building façade, Riverside, 58
Wright, Frank Lloyd, 31, 68
Wright, Lloyd, 120

YWCA, Riverside (Morgan), 57, *58*, 78, 91–92, *92*, 118

Zurbarán, Francisco de, 85

This book is published in conjunction with the exhibition *Myth & Mirage: Inland Southern California, Birthplace of the Spanish Colonial Revival*, presented at the Riverside Art Museum from September 22, 2017, to January 28, 2018.

Pacific Standard Time: LA/LA
Latin American & Latino Art in LA

Presenting Sponsors
The Getty
Bank of America

Major support for this exhibition and publication is provided through grants from the Getty Foundation.

Myth & Mirage: Inland Southern California, Birthplace of the Spanish Colonial Revival is part of Pacific Standard Time: LA/LA, a far-reaching and ambitious exploration of Latin American and Latino art in dialogue with Los Angeles, taking place from September 2017 through January 2018 at more than 70 cultural institutions across Southern California. Pacific Standard Time is an initiative of the Getty.

COPYRIGHT © 2017

All rights reserved. No part of this publication may be reproduced or transmitted in any form or by any means, electronic or mechanical, including photocopy, recording, or any information storage or retrieval system, without permission in writing from the publisher.

LIBRARY OF CONGRESS CATALOGING-IN-PUBLICATION DATA

Betsky, Aaron. | McCulloh, Douglas, photographer (expression) | Riverside Art Museum. organizer, host institution.

Title: Myth and mirage : inland Southern California, birthplace of the Spanish colonial revival / essays by Aaron Betsky, H. Vincent Moses and Catherine Whitmore, Lindsey Rossi, Carolyn Schutten, Patricia Morton, Ronald Ellis (Interview), Susan Straight ; photographs by Douglas McCulloh.

Description: Riverside, California : Riverside Art Museum, 2017. | "This book is published in conjunction with the exhibition Myth & Mirage: Inland Southern California, Birthplace of the Spanish Colonial Revival, presented at the Riverside Art Museum from September 22, 2017 to January 28, 2018." | Includes bibliographical references and index.

Identifiers: LCCN 2017016062 | ISBN 9780980220766 (hardcover : alk. paper)

Subjects: LCSH: Colonial revival (Architecture)—California, Southern—Exhibitions. | Architecture, Spanish colonial—Influence—Exhibitions. | Symbolism in architecture—California, Southern—Exhibitions.

Classification: LCC NA730.C22 S686 2017 | DDC 720.9794/9—dc23

LC record available at https://lccn.loc.gov/2017016062

PUBLISHED BY THE RIVERSIDE ART MUSEUM
3425 Mission Inn Avenue
Riverside, California 92501
www.riversideartmuseum.org

AVAILABLE THROUGH:
ARTBOOK | D.A.P.
75 Broad Street, Suite 630
New York, NY 10004
www.artbook.com

PRODUCED BY LUCIA | MARQUAND, SEATTLE
www.luciamarquand.com

Edited by Melissa Duffes
Designed by Thomas Eykemans
Typeset in Adobe Caslon by Integrated Composition Systems
Proofread by Barbara Bowen
Indexed by David Luljak
Color management by iocolor, Seattle
Printed and bound in China by Artron Art Group

IMAGE CREDITS

All photos by Douglas McCulloh unless otherwise noted. Corona Public Library History Room: p. 61, figs. 20, 22. Digital Public Library of America: p. 65, fig. 26; http://boingboing.net/2014/06/11/1949-chart-shows-difference-be.html: p. 136, fig. 7; Mission Inn Foundation Museum Archive: p. 84, fig. 9; p. 88, fig. 14; p. 100, fig. 1; p. 168, figs. 5–6; p. 176, fig. 19; Drew Oberjuerge Collection: p. 79, fig. 1; p. 80, fig. 2; p. 83, fig. 8; p. 87, figs. 12–13; Charles Phoenix Collection: p. 18, fig. 12; Riverside Metropolitan Museum Archive: p. 15, figs. 9–10; p. 44, fig. 1; p. 45, fig. 2; p. 46, fig. 3; p. 47, fig. 4; p. 50, fig. 7; p. 57, fig. 14; p. 58, fig. 16; p. 62, figs. 23–23a; p. 63, fig. 24; p. 65, fig. 25; p. 102, figs. 2–3; p. 103, fig. 4; p. 117, fig. 6; p. 118, fig. 7; p. 121, fig. 10; p. 122, fig. 11; p. 123, fig. 12; p. 136, fig. 6; Riverside Public Library Local History Resource Center: p. 50, fig. 6; A. K. Smiley Public Library History Room: p. 51, fig. 8; p. 53, fig. 10; p. 67, fig. 29; p. 113, fig. 2; p. 114, fig. 3; p. 119, fig. 8; Taco Bell Corporate Archives: p. 143, fig. 14; Used by permission of Special Collections & University Archives, UCR Library, University of California, Riverside: p. 105, fig. 6; p. 112, fig. 1; p. 115, fig. 4; p. 116, fig. 5; p. 120, fig. 9; Tom Zimmerman Collection: p. 15, fig. 8; p. 16, fig. 11